ENVIRONMENTAL CONSCIENCE
Socio-Legal and Judicial Paradigm

About the Editors

Manmohan Singh Gill is Professor in the Department of Sociology, Guru Nanak Dev University, Amritsar since 1986. He has served in the capacity of Reader, Academic Staff College and Programme Coordinator, NSS in the same university.

With 24 years of teaching and research experience, Dr. Gill has specialized in the fields of Rural Sociology and Social Ecology. He has completed research projects of the Department of Social Welfare of Punjab Government on Untouchability and UGC project on Caste-Land Relations. He has published numerous papers in research journals, besides making presentations in various academic forums at national and international levels. He has authored books on *Spatial Patterns: Socio-ecological Analysis* (1991) and *Punjab Society: Perspectives and Challenges* (2003).

Dr. Gill is life member of Indian Sociological Society and North-West Sociological Association. He has edited Guru Nanak Journal of Sociology for three years (2004-2006). For his service to society in the field of environment, he has been awarded Vijay Rattan for the year 2005. He is on the advisory body of All India Foundation for Peace and Disaster Management (New Delhi).

Jasleen Kewlani is faculty in Sociology and also holding the charge of Deputy Coordinator of RGNUL Institute for Competitive Examination (RICE). Prior to joining RGNUL, she had taught in S.D. Arya College, Dinanagar; Lovely Institute of Law, Phagwara, and Guru Nanak Dev University, Amritsar. Her areas of expertise are Environmental Sociology, Rural and Urban Studies, and Women Studies. She has presented (about 30) research papers in numerous seminars, workshops and conferences. In the year 2003, the USA Human Rights Campaign awarded her scholarship for two years at USA for her endeavour for human rights in India.

Environmental Conscience
Socio-Legal and Judicial Paradigm

Edited by
Manmohan Singh Gill
Jasleen Kewlani

CONCEPT PUBLISHING COMPANY, NEW DELHI-110059

ISBN-13: 978-81-8069-601-5 **ISBN-10: 81-8069-601-4**

First Published 2009

Published and Printed by

Concept
Since 1974

Concept Publishing Company
Regd. Office:
A/15-16, Commercial Block, Mohan Garden
New Delhi-110059 (India)
Phones : 25351460, 25351794, *Fax* : 091-11-25357109
Email : publishing@conceptpub.com,
website: www.conceptpub.com

Editorial Office:
H-13, Bali Nagar, New Delhi-110 015, India.

Cataloging in Publication Data--*Courtesy:* D.K. Agencies (P) Ltd. <docinfo@dkagencies.com>

Environmental conscience : socio-legal and judicial paradigm / edited by Manmohan Singh Gill, Jasleen Kewlani.
p. cm.
Contributed articles.
In Indian context.
Includes bibliographical references and index.
ISBN 13: 9788180696015 ISBN 10: 8180696014

1. Human ecology--India. 2. Economic development--Environmental aspects--India. 3. Environmental degradation--India. 4. Environmental protection--India. 5. Environmental law--India. 6. Sustainable development. I. Gill, Manmohan Singh, 1952- II. Kewlani, Jasleen.

DDC 363.709 54 22

R.L. BHATIA

GOVERNOR OF KERALA

KERALA RAJ BHAVAN
THIRUVANANTHAPURAM
PIN 695 099

Foreword

Ours is an era when economic development overtakes all other cores of survival such as ethics, values, traditions, religion, environment, etc. And, as such, attaining sustainable development seems to be a far-off target. The pros and cons of this issue have been effectively dealt with in 'Environment Conscience'. The various papers incorporated in it provide insight into the main theme as a workable instrument for an ecologically sustainable society. Moving ahead of the commonly discussed definitions and sources, the contributors attempt a broader exploration of the environmental issues, while the recommendations they putforth form a sound base for policy-making.

The role of the Judiciary and the Legislature in the context of environmental management and protection opens new zones for further research. Various case laws referred in the papers elaborate the environmental conscience of the judiciary which further justifies the title of the book. Another commendable feature is the special emphasis on the rural and the tribal folks. The environment is of utmost importance to all the people, but it is more so in the case of the rural and tribal folks who cannot afford the artificially created sources of livelihood needs and hence largely depend upon the natural sources in their natural environment.

The papers on Social and Judicial Activism, Socio-Cultural Issues related to Sustainability, and Impact of growth of population on Environment can be a workable source of amendments in the existing law to make it further successful. Meanwhile, international comparisons and references made in lieu of environmental degradation make the book interesting and capable of guiding the communities of academicians and researchers.

I hope this volume will be a great source of mobilization for the governmental and non-governmental bodies working for environmental cause, making them more fruitfully oriented for the purpose of protection, preservation and upgradation of the environment. 'Environmental Conscience', I am sure, will be a success prompting its editors to embark on similar enterprises in future as well.

R.L. Bhatia

Introduction

The general understanding about the preservation and management of natural resources and environmental issues came to the fore after various deliberations on environmental issues taken up at the international forums. The problem is the outcome of direct or indirect consequence of human activities blatantly progressing towards economic development. The objective of growth has shifted from one form of development to another but even the sustainable development path treaded to meet the demands of the future generation falters in the realm of struggle to gain control over the resources covertly or overtly. The advancement in technology, new discoveries, and innovations has enhanced the problem rather than tackling the issue in its right spirit. There is greater misuse of the resources as well as technologies in the hands of populace and nations abreasted with the slogan of consumerism. The increase in the stock of e-waste and plastic polymers has further aggravated the problem. The developed nations are finding developing nations as their dumping grounds for the e-waste and other such materials. The world oceans are accumulating immense amount of plastic debris which is not biodegradable and effect the marine food chain. Further, conflicts over natural resources have intensified in many parts of the world. With the present pace of population growth, more so in developing countries, and the demand on the resources, conflicts and struggle over the use of resources contemplate into violent struggle among the communities, thereby posing a serious threat to peace and stability of human society.

Realization about the path of development and inequalities widening among the nations and people, environment has been taken as a human rights issue and many nations have incorporated this as Fundamental Right as in India. There has been increasing involvement of Non-governmental Organizations in environmental discourse. Peoples' participation is considered as a regulatory policy in management of forests with all its limitations. At the same time, natural resources cannot be preserved and managed without the proactive policies of States and participation of judiciary, people and Non-Governmental Organizations. Moreover, there is a callous attitude towards waste generation and its disposal by the people and the organizations entrusted such responsibility. This neglect and indifference to a large extent is the outcome of lack of knowledge and awareness about the threat imposed by the consumptions of resources and generation of waste and its management. Not only this, the people are remorsely concerned with the protection, management and improvement of natural resources including forests, lakes, rivers etc. Facts such as these force us to confront vital questions: what are the theoretical frameworks within which the environmental issues can be understood? Is it the poverty, population, governance and the like that is the missing link to understand the problem and realization of its gravity? Is the sustainable goal of development achievable? What is the role of judiciary in the whole process? Surely, there is no better way to begin the process of answering these questions through such a reading. It is the need of the hour to create a consciousness among the masses about environmental issues in their logistic understanding and in the light of globalization associated with economic boon and consumerism connotation that usurp the vision of even the enlightened man. This reader touches the issue in such a logical manner that flows from the theoretical understanding to realization of action at different ends to meet the objective of this reader.

Organization of the Reader

The work is organized in a manner that depicts the theoretical frame of the problem and then explores the different dimensions of the issue associated with the realization of environment as a cause and consequence of human activities. Clearly differentiating 'shallow' and 'deep' ecological perspectives, the essay on "Ecology and Environment: Theoretical Orientation" lays emphasis on cultural, religious and social values that relates man to his environment and a social binding to protect the environment. In her essay, Jasleen has tried to apply system theory to the process of development and legislative procedure. For the author, theoretical perspectives and their orientation have potential to analyze and balance the internalities and externalities of economy, environment and ecology. Development process has generated inequalities among the nations with stress on the developing countries, which are already under duress due to their population explosion. These countries are depriving communities of their natural habitat and vital subsistence resources to make eco-material available for their growing population. Analyzing this aspect, Gill in his paper 'Impact of Development and Population on Environment', states that lack of realization on the part of developed and affluent sections of the nations and people about the under-privileged and under-developed nations is missing and this is the basic cause of environmental degradation.

Environmental degradation has systematically and cumulatively changed the environment. Countries like USA, Canada, Netherland and others in their race to economic development have brought about economic disparities at regional and national levels. The environmental problem in these countries may vary but its tragic disasters are looming equally over all the nations of the world. The paper "Global Environmental Risk: A Threat to Sustainability" refers to creating a social vision with discourse on environmental challenges so as to generate consciousness among the participating nations to become open and permeate information and technological advancement to combat global environmental threats and sustainability of society. Sustainability of society lies in exploring the rich tradition of the developing nations, which has accumulated their cultural heritage with direct interaction between nature and man. To Gill, "Sustainable Society: Inhibitions and Facilitations", beliefs, culture and value patterns of society; social processes; the economic differentiation of various societies and groups etc. forward such complexities, which pose numerous hindrances in the way of sustainable growth of society. He refers to a new social order acting as a cause and consequence of many socio-cultural alterations. While exploring the "Problems of Sustainability in Third World Countries", Gill has referred to certain features of society that inhibit the mechanism of sustainability. The author has incorporated a detail discourse on regional issues and circumstances that need a conscious and concerted effort through enlightenment of self.

Nevertheless, science and technology are the cause and consequence of environmental changes; their development renders deprivation and restrains sustainability of society. Survival of human life and its security is constrained by environmental changes and deprivation of natural resources. Science and technology has trapped humanity in its dichotomous cover of survival and adversity. Realizing this need, Jasleen in her essay entitled "Development of Science and Technology: Impact on Human Security and Survival" has explained various dimensions of the impact through secondary sources. She wants to arouse a consciousness that development of science and technology should not surpass community sentiments and community needs. In the same vain, she and Sharma, have discussed the formulation of economic policies and environmental consequences. In the paper, "Economic Policy Making and Environmental Consequences", Jasleen and Sharma points out that society cannot sustain resources if their consumption cut into the standing stock of ecological resources. The real challenge lies in taking advantages of existing technology and knowledge to optimize quality of life rather than maximizing the quantity of material life. They pose a question on the veracity of WTO towards sustainability of society and environmental issues.

The reader after making observation on the theoretical issues and some of the important variables that have consequential impact on the environment, ecology and sustainability, its focus shifted to empirical situation. Chatterjee, "Appraising a Local Environmental Movement in India", finds that a number of movements have been initiated to arouse the environmental issues at national level, some time taking the shape of international movement. These movements have social, economic and political dimensions but local issues have been neglected to a large extent. Based on data collected from the villages of West Bengal, the author sways on the issuc that led to the emergence of a local movement against air pollution caused by thermal plant. He suggests that scientists, natural and social together; find out a balance between development and environment otherwisc people will create a situation where environmental consequences of development activities could hardly be ignored. Similarly, Kewlani and Gill have discussed the situation of urban areas with regard to its environmental problems and governance to mitigate these. Their analysis is based on Census of India, 2001. Overcrowding of urban areas, stress on urban infrastructure and its resources have been explained. These problems can be tackled with the active participation of local bodies, NGOs and the people. Another issue involves the wealth of the mother earth that is the forests. Deforestation has not only destroyed the local communities' sustenance base but also the bio-diversity of the area. Forests are the mechanism for sustainability and any attempt of deforestation need to be evaluated in terms of 'Environmental cost benefit analysis'. This has been the crux of the essay on "Environmental Degeneration with focused reference to Deforestation..." by Kewlani. Further, any mass level campaign needs the contribution of involved communities. Even for deriving solutions to the environmental problems and their causative influxes, awareness of the masses contributes considerably in their mobilization and redressal. Gill and Kaur in their paper, "Environmental Management: a Study in Environmental Awareness" have made an attempt to explore the limits of environmental awareness amongst the school going children. Based on empirical data, they have generalized their finding for the public and private schools.

With greater realization of the environmental changes and their consequential impact, people's activism encourages large number of litigations: which further design a laboratory to measure the environmental consciousness of the judiciary. In their paper, "Environmental Laws and the Indian Judiciary", Kewlani and Sharma have incorporated various landmark cases that came up during 2007-2008. These cases deal with multiple angles of environmental concern and elaborate the role of judiciary in maintaining and protecting environment. The judgments derived by the courts act as facilitation and some where as inhibition for ecological interest of the nation. Kanwar and Singla have emphasized on judicial activism as a moral force encouraged/being encouraged by the concept of Public Interest Litigations (PIL). The paper refers to a legal lens which judiciary has applied to account environmental laws with human rights. On the other hand, Bansal and Kalra have analyzed the role of judiciary in the management, protection and preservation of environment in their paper "Role of Judiciary in Environmental Activism". With all pervasive realization and active involvement of judiciary, ecological degradation continues unabatedly. The authors have suggested some workable solutions to control declining environmental standards of the contemporary technologically advancing India. Gigimon rather, in his essay "Humanitarian Faces and Legal Hues of Aura", concludes that the most important instrument to preserve environment is to redefine the rubrics of development. In spite of relatively well-defined relation between environmental protection and human rights protection, the issue suffers from inattentiveness in human rights movements as well as environmental movements.

The reader though explores a wide range of environmental themes through various ecological perspectives, further encapsulating population, technology, economy, law making, adjudication, it shall be an instrument for understanding the complexities entrenched in legislative and jurisprudence with reference to environment. This will sensitize the people about the causes and consequential effects of environment. Going through the wide range of topics covered under this reader, a number of hypotheses can be formulated for initiating further research in environment and ecology. It is inevitable that certain other valuable issues have been left out. Such is the nature of a reader of this sort.

Contents

Acknowledgements

The present work is an outcome of series of discussions with teachers, students and researchers about the degrading situation of the environment, a symbiotic relation between man and nature and conflict over resources. This book would not have been possible without the sincere support of various researchers who contributed their papers with sincerity to the cause of environmental awareness. The present book represents a tentative framework of the essential reading that is prerequisite to enlighten the masses on environmental paradigms. We shared this judgment with colleagues and members from faculty of environmental sciences and law. All the contributors deserve special thanks that enriched the discourse on environmental conscience.

We are highly indebted to His Excellency Sh. R.L. Bhatia, the Governor of Kerala, who has gone through the draft with keen interest despite his busy schedule at Raj Bhavan and contributing the Foreword of this reader.

We commend the efforts of the Proprietor, Sh. A.K. Mittal and the staff of Concept Publishing Company, New Delhi who helped to bring to fruition the idea of this book.

This work owes much to Mrs. Ravinder, Mr. and Mrs. Surinder Kewlani and many others for their cooperation, encouragement and suggestions to complete this work.

Manmohan S. Gill
Professor, Department of Sociology
Guru Nanak Dev University
Amritsar

Jasleen Kewlani
Lecturer in Sociology
Rajiv Gandhi National University of Law
Patiala

List of Contributors

Chandan Singla: As a student of Rajiv Gandhi National University of Law, Patiala, Punjab, he has visited Europe to participate in the 15th Willian C. Vis International Commercial Arbitration International Moot Court Competition, Vienna, Austria.

Deba Prashad Chatterjee: He is a Reader at the Department of Sociology, Maulana Azad College, the University of Calcutta, West Bengal, India. As an environmental sociologist, he has recently completed a Public Health and Engineering, Government of West Bengal funded project on the Provision of Safe Drinking Water in Some Rural Areas of West Bengal. At present, he is working as a country co-investigator for India in a cross-cultural research project titled, 'Collaborative Research: Advocacy Coalitions as Agents of Change in Climate Change Policy-Making' under the leadership of Professor Jeffrey Broadbent of the University of Minnesota, USA. His other areas of interest include population studies, sociological theory and thought, social movement and Indian sociology. Recently he has published a chapter entitled, 'Sociology in Different Functional Aspects of Society and Knowledge' in an edited volume ('Sociology in Practice—An Introduction', edited by Debamitra), published by the ICFAI University Press, Hyderabad, India.

Gigimon V.S.: He is working as Lecturer in Rajiv Gandhi National University of Law for the last two years. He has many publications to his credit particularly on Environmental Laws and Human Rights.

Jasleen Kewlani: She is working as lecturer in Sociology and is also holding the charge of Deputy Coordinator of RGNUL Institute for Competitive Examination (RICE). Prior to joining RGNUL, she has taught at S.D. Arya Mahila College, Dinanagar and Guru Nanak Dev University, Amritsar. Her areas of expertise are Environmental Sociology, Rural and Urban Studies, and Women Studies. She has also organized various national as well as international events related to academics and extra co-curricular activities. She is also a member of various national organizations and associations dealing in academics and social welfare.

Karan Kalra: He is a student of B.A.LL.B. in Rajiv Gandhi National University of Law, Patiala, Punjab. He participated in the 5th Willian C. Vis (East) International Commercial Arbitration Moot, held in Hong Kong. He has made presentation of papers in national as well as international seminars and conferences.

Manmohan S. Gill: He is professor of Sociology at Guru Nanak Dev University, Amritsar. He has a teaching experience of 24 years and is associated with many academic bodies of the university. His areas of specialization are Rural Sociology and Urban Ecology. He has completed research projects on Untouchability and Dynamics of Caste-Land Relations. He has two books to his credit: *Punjab Society* (2003) and *Spatial Patterns* (1991). He is the recipient of Vijay Rattan award for his outstanding work in the field of environmental education for the year 2005. He is on the advisory panel of All India Foundation for Peace and Disaster Management (New Delhi).

Navdeep Kaur: She has been a research student in Department of Sociology, Guru Nanak Dev University, Amritsar and has worked on the project Environmental Awareness.

Nishant Sharma: He is a student of Law in Rajiv Gandhi National University of Law, Patiala, Punjab. He has authored about six research papers and contributed them in national as well as international conferences and seminars.

Prashish Kanwar: He works with Rajiv Gandhi National University of Law, Patiala, Punjab. He has been involved in research related activities like writing research papers and handling empirical research in internship programmes also. He has got his papers published in national as well as international referred journals and edited books too. In the recent times, he visited Europe to participate in the 15th Willian C. Vis International Commercial Arbitration International Moot Court Competition, Vienna, Austria. He has also been recently selected to attend the International Youth Leadership Conference, being held at Prague.

Rajan Bansal: He is a student of B.A. LL.B. in Rajiv Gandhi National University of Law, Patiala, Punjab and has participated in the 5th William C. Vis (East) International Commercial Arbitration Moot, held in Hong Kong. He has presented papers in national as well as international seminars and conferences.

1

Ecology and Environment
Theoretical Orientations

JASLEEN KEWLANI

Introduction

The increasing realization of the vitality of environment and its issues, has led to emergence of 'environmentalism', "which means ideologies and practices which inform and flow from a concern with environment" (Pepper, 1986, p.13). In context of contemporary environment crisis, environmental science takes a prenatal position, since it deals with the application of knowledge gained with the help of ecology for managing environment, along with its various off-shoots. Environment can be natural as well as man-made. Natural environment consists of air, water, soil, land, forests, wildlife, etc. and man-made environment which includes housing, technology, transportation, settlement, utilities, etc. To know, what type of link exists between these two forms of environment and what are the effects of this inter-linkage, (positive or negative), is one of the major aims of this research? The environmental studies are very much related with Botany, Zoology, Chemistry, Pedology, Geography, Climatology, Oceanography, Geomorphology, Sociology, Political Science, Economics, Management, etc. Since environment and man are the primary concerns of environmental studies, all the subjects related with these aspects have relationship 'Controller' of life and living styles of human beings (Saxena, 2006). Environmental studies are the need of the hour because with an indepth understanding of environment; firstly, various facets and problems of survival are interpreted in authentic terms and secondly, a strategic planning can be made. To have complete analysis of environment and ecology there is a need to refer to various ecological perspectives, which have been given by prominent environmentalists, sociologists and geologists. These ecological perspectives forward sundry approaches which have been used till date for rendering ecological analysis or environmental studies.

Ecological Perspectives

While analyzing ecological issues, two perspectives came forward; (1) "Reformist Ecological Perspective". Reformist ecological perspective is also known by the name of "Shallow Ecological Perspective"; and (2) "Deep Ecological Perspective". The difference between these two perspectives, didn't lie in the 'ends' but in the 'means' which were adopted to achieve those ends. Though target of both of these ecological perspectives is environmental protection, but the ideology and means to manage or reconcile environmental destruction are distinct. The reformist ecology targets to reform and control the worst which has gone with the environmental entities of nature, i.e. air, water, land, etc. It refers to several social movements, which were a product of man's quest to improve the degenerating environmental quality. Environment has always been considered an important part of one's life, hence the goal of all such movements initiated by reformist ecologists, was to change society for 'better living'. Along with the motive of changing environment's scene, criteria for survival remained intact. According to Reformists, any alternate process must not attack the premise of the "dominant social paradigm", which expresses the mental image of social reality that guides expectation in society. This dominant social paradigm, according to reformist ecologists consists of basic values, culture and heritage of masses, which is deep-rooted in their social survival entangled with their relative environmental circumstances and ecological surroundings. Hence, they believe that environment must improve but it should not be accompanied with any alteration in basic dominant socio-cultural paradigm of individuals (Merchant, 1996).

On the other hand, "Deep Ecological Perspective" forwards an altered mode of discussion, posing an absolutely contrary notion to that of the Reformist Ecological Perspective. The norms and tendencies of the Deep Ecology Movements are not derived calculatively from ecology or induction, rather from the knowledge and attributes gained by the researchers working in ecological domain. The contents and their understanding require an ability to construct a conserver society. When Reformist Ecology failed to protect environment and ecology from further deterioration, Deep Ecological Perspective began functionally crucial. Arne Naess (1989) was the founder of this school of ecological thought who thinks that 'Environmental Crisis' is the 'Crises of Consciousness'. Arne Naess has been an environmentalist who stressed on "Ecological Egalitarianism' as a reflexive component of Deep Ecological Perspective. Ecological Egalitarianism implied that all individuals have equal rights to live and blossom in an intuitively clear and obvious value axiom. This assertion supported that Deep Ecology is forcefully normative in nature. It is ecophilosophical not mere ecological. In addition to this view, Naess talks of Principles of 'Diversity' and 'Symboisis'. According to him, 'Diversity' enhances the potentialities of survival and the chores of new modes of life; and 'Symbiosis' tends to reduce the multiplicity of kinds of forms of life and create integration within the communities of the same species. This helps to maintain equilibrium which glibs the survival of individuals. Infact Deep Ecology and its supporters support the fight against economic, cultural, military invasion and domination which leads to degradation of environmental quality.

The thought given by Naess was strengthened by Eckersley (1995), who added dimension of social justice and human emancipation along with a change in the way of thinking about the natural world. These ideas were further strengthened by Carlos Castaneda. Carlos Castaneda (1971) says that it is a challenging task for modern man to acquire and develop understanding of Deep Ecology because it requires a major perceptual shift of man's nature. Further, Deep Ecology gained popularity in Native Americans during 1960-1970. These Native Americans had 'separate realities' which were entangled to their life style and closely linked with environment around them. In fact Castaneda aimed at understanding how did these tribes of Native Americans cope with the changes occurring due to technological innovations and advancements in their natural environment. According to him, the modern western man cannot understand these 'separate realities', which owe their origin to the basic no.ms of the tribes. In reaction to Castaneda views about Deep Ecological Perspective, Robert Orustein (1996) forwarded his views that the Western trained and even those who are a 'seeker' are by their culture almost unprepared to understand the esoteric traditions constituting the separate realities. This counter response shows that Castaneda's work on Deep Ecology had been critically appraised due to his undermining of the modern man's perceptual ability to interpret the importance of environment.

Another imminent component of Deep Ecology is Ramachandra Guha (1996) who stated that Deep Ecology shows man's arrogance and his ecological hubris, which puts environment at the top of apex of life choices. This perspective according to Guha depicts a transition from anthropocentric to a biocentric approach of man. Anthropocentric approach is a 'human' centred approach and bio-centric approach interprets human as only one of the elements in the ecosystems but doesn't attribute indispensability to his existence. In simple terms, anthropocentric approach gives importance to human as an element of study but biocentric approach gives cardinal emphasise to ecology and environment. According to Guha, any form of intervention in nature and its paradigm should be guided primarily by the need to preserve biotic integrity rather than by the needs of humans: Guha upheld a statement that Deep Ecology encourages political militancy not negotiations and advances philosophical arguments not aesthetic notions. This statement infact makes the comparison between reformist and Deep Ecological Perspective more ocular. In simple terms, Deep Ecological Perspective is fundamentally normative in nature, providing importance to ethics determing the attitude of individuals. According to Pojman (2000), if environment is to be protected, a deep-rooted, normative as well as an attitudinal approach is to be adopted, which ought to be objective in nature.

Another sociologist named Bill Devall (1996) further elaborated clearly the contents of Deep Ecological Perspective. According to him, usable instruments of Deep Ecological Perspectives are local autonomy, decentralization of power, steady scale economy and a radically changed social structure. The word 'radical' was specifically stressed by Devall. He provided alternate terminology for 'Deep Ecology'. He used some terms like 'Radical Sociology' and

'Revolutionary Ecology'. He upheld that Deep Ecology Perspective was a reaction to the successes and excesses of the implementation of the dominant social paradigm and its assumptions. This 'dominant social paradigm' reflects mental image of social reality that guides the expectation in a society. The Reformist Ecology didn't intend to alter the dominant social paradigm which was attacked by normative strategy of Deep Ecology later. The 'Dominant Social Paradigm' comprises of the social fabrics including kinship and family structure; naturalistic tendencies of a society; and modes of adapting to the recently occurring social changes. Deep Ecological Perspective aims at improving environmental quality and checking further degradation of environment by attacking the fundamental nature of this paradigm, because it needs a shift of normative aspect of human attitude. When environmental degradation has reached a drastic extent, an abrupt and a strong action is needed to manage it. Keeping this assertion in view, Deep Ecology becomes the most functional tool to formulate the required management strategy. In all, a stress is laid to understand those cultural, religious and social values that related man to his environment and made a social binding on the individual to protect the environment. The major emphasis will be on empowerment, social justice, class differentiation and struggle against resource depletion and environmental pollution. As Deep Ecology encompasses various aspects of life, it is important to discuss holistic approach.

Other than the above discussed ecological perspectives, another theory has been very prominent in environmental studies, i.e. "System Theory". Moving towards the importance of a holistic approach to analysis; 'System Theory' is of a great relevance for environmental studies. The conception of development has changed and the new concept of development as a process incorporates various fields of life together. Living organisms cover a wide range of phenomenon including individual organisms, social systems and ecosystems. Hence systems theory becomes vital. "Common framework and language for biology, psychology, medicine, economics, ecology and many other sciences, a framework in which the so urgently needed ecological perspective is explicitly manifest" (Merchant, 1996). System theory assumes that all the aspects of life together formulate a system, which is to be regularized in order to attain an 'all over development'. This theory implies shift (i) from part to the whole,

(ii) from structure to the process of maintenance, (iii) from objective to epistemic science; which aims at clearly defining the ecological concepts, (iv) from "building" to "network" as a metaphor of knowledge, and (v) from truth to approximate description, which reinforces human dominance as an 'observer' who identifies various objects of nature and describes them. Imminent supporters and contributors of the theory are: Heisenberg (1971), who wrote a book titled "Physics and Philosophy" and his another book titled "Der Teil Und Ganze" (1970), states that nature is also exposed to questions. Capra (1985) while taking an interview of Geoffrey Chew, reproduced Chew's views in one of his intellectual productions and explains Chew's idea that understanding of nature needs to be undertaken entirely through self-consistency, because environment as a subject cannot be reduced to any fundamental entities. There are no fundamental equations or fundamental symmetries in environmental understanding. Environment as a physical reality is seen as a dynamic web of interrelated events related to socio-cultural systems as well. Therefore, these interrelated events need a cognitive understanding for a absolute interpretation of the social world. Thomas Kuhn (1970) also finds system theory also of great importance and states in his theory of Scientific Revolution that changes of Paradigms occur in discontinuous, revolutionary breaks, called 'paradigm shifts'. Keeping this alternation in view, it can be said that the modern world recognizes, paradigm shift in physics as an integral part of a much larger cultural transformation. Einstein Albert (1949) has also presumed a similar view to Kuhn that no fundamental theoretical foundation exist in physics and rich its understanding. Hence, an amalgamated approach is needed which is catered by System Theory. The motive of explaining "system theory" and its components is not to add one more theoretical perspective, but to reaffirm that all the aspects of life are inter-woven and make a system in all. This research aims at analyzing that how far this System Theory has been accepted and applied and how far has it been internalized in the process of development and legislative procedure, along with the effects in case of non-adoption of a system approach in the procession towards growth.

Theorists like Boserup further produced theories on environment with 'Gender Dimension' as the cardinal emphasis. This theory further forwarded a novel impetus to the already existing theories on Ecology; because prior to it, there was no theory dealing with environment or any other allied concept through the gender dimension. Boserup in 1970 produced a work titled, "Women's role in economic development", and held that women are an imminent part of gendered division of labour, participating in economic development. This book was considered as a catalyst for the growth of a body of literature on Women's role in development. In addition to Boserup's theory, Brenda Cranney (2001) stressed upon one aspect of social reproduction, which is sexual or social division of labour in home and in farm work. Sexual division of labour refers to a division created by cultural and social norms and not by biological difference. An encapsulation of research findings show that all societies develop a sexual division of labour that varies across countries, cultures and history, thereby implying that there is nothing natural or biological in such division (Beneria and Roldan 1987; Kelkar and Nathan 1991; Rogers 1981; Rubin 1975). On the other hand, Cranney aimed at analyzing women's contribution to production and social reproduction, which is necessary to determine what labour women perform for the family and how and why specific tasks are allocated. Both in primitive and in more developed communities, traditional division of labour in both farms and families, is usually considered natural, being sexually determined. The persons of a specific community may feel that in other communities, there exist different ways of determining division of labour (dividing the work burden among sexes), which even they state as 'natural' basis of division of labour (Boserup, 1970). Anyhow, women's work does not happen in a vacuum. It is influenced by capitalistic development, which in turn influences the dynamics of family life. Linking all these thoughts and facts with the main theme of the research, it is evident that environmental degradation affects females, their livelihood and their household chores up to a large extent. Specifically feminist context, deforestation affects the availability of the forest products like fuel wood, food, etc. and also the time required to collect these products. This very phenomenon increases the work load for women and decreases household nutrition (Venkatestwara, 1995). A study carried out by Cranney, in mountain region of Himachal Pradesh proves that deforestation doubles the women's workload. The women of this area identify water and the forests as two single most important factors in their daily work and life that consumes their most of the time. They still consider water supply as a secondary issue to forests as in falls, winters and early springs, their water needs are met but the forests affect every aspect of their life everyday. Women collect two types of wood; for fodder and for fuel consumption. Fodder consists of branches of specific trees that produce leaves suitable for animals to eat. Fodder is usually collected twice a day as the leaves need to be fresh for the cows to eat. The branches are once thrown in the cow yards and after the leaves are eaten, the twigs are used for fuelwood in chullah (Cranney Brenda, 2001).

She further states that women go and collect these products from forests and when deforestation occurs women's burden of work gets intensified in terms of time and energy. A common complaint of women of Himachal Pradesh has been that they had to travel from 1 to 10 kilometres to collect wood. The weight of the load varies from 10 to 40 kilometres which is not in appropriate connotation with each woman's stamina and potential. Eventually if not immediately, the incident has alarming effects on women's general health latently. Not only health, but education which is the strongest wheel of future development of women, also suffers. Girl child is eliminated from her school and is imposed upon to share the household burden of the mothers because after deforestation they spend about 8-10 hours a day on walking to the forests and coming back after collecting the forest products required for household maintenance. Another content to be considered in context of women's relation with development and her importance in growth paradigm is 'Ecofeminism'. Contributions have been made by liberal, socialist and radical ecofeminists. Liberal ecofeminists aim at redistributive policy changes and do not at all recommend a full-fledged restructuring of the whole set ups. On the other hand, socialist feminists want equal representation of women in government and non-governmental organisation (Berman, 1993). Another ecofeminist perspective is Radical ecofeminism which asserts that women have a spiritual and biological link to the non-human world and of course are closer to nature posing a 'divinity factor', as compared to other gender.

Theoretical perspectives on ecology reflect a mixture of crucial zones of survival. As these theories include psychological, economical, social, cultural and even gendered aspects of interpretation. Reformist Ecological Perspective and Deep Ecological Perspectives pose entirely different domains of thoughts for the same motive. The followers of these two groups though possessed same object and similar objectives; but still the ideologies rendering achievement strategies were different and diverse. Environmental protection, preservation and management have been defined as the main goal of these perspectives but the intensity of incorporation of the goal varies. These two ecological perspectives reflect psychological and ideological stand of the related communities. In addition to the discussion, 'System Theory' incorporated in the preceding part of the paper caters to even higher onus. It produces a higher 'social responsibility' and 'academic obligation' by talking about a 'Common Framework'. "Common Framework" and language for biology, psychology, medicine, economics, ecology and many other sciences; a framework in which the so urgently needed ecological perspective is explicitly manifest. This theory makes 'Environment' a commonly shared responsibility of all the disciplines of knowledge. Another Theoretical Perspective forwarded by Boserup and then empirically endorsed by Cranney Brenda exposes ecology and ecological studies to Gender dimensions. This theoretical perspective gave a new turn to the ecological thought by incorporating feminine dimensions into it. Not only this, a further differentiation has been mentioned as to how various segments of the same gender interpret their ecological internalities. All the ecological perspectives added hereby encapsulate almost all the dimensions of survival. But the analysis of the existent theoretical perspectives on ecology in modern world needs a complementation and confrontation with the theories of globalization. Ecological degradation has aggravated since the past; and the major reason is universally increasing greed for economic development, which is further a survival criterion for the globalizing and the globalized world. The coming part of the paper focuses on elucidating leading theoretical approaches to the understanding and interpretations of globalization, in both its current form and potential future shapes. The debates and discussion on environmental degradation started catching 'intensive' as well 'extensive' concern of academicians, environmentalists, economists, scientists and even common masses only when globalization became the call of the world. In rapidly globalizing world, all the nations are marching ahead with an economic zeal to be called 'successful industrialized economies'. The reason is that industrialization is taken as the most potential instrument of economic development. It of course leads to further downfall of ecological standards, though it reflects one-sided development, i.e. economic development. Certainly, economic development is the need of the hour, without which handling the consistently growing population is not possible. But, the processes and the procedures undertaken for the motive of economic development do not reflect the feature of sustainability; due to which environment is being affected in an unhesitant manner.

The era of globalization having a quest for economic development has given new term and essence to theory also. Environment was being utilized in the past and further it started getting exploited wantonly; and now it is 'attacked'. The attack on environment and its units is motivated by the need of economic growth and the related objectives; but it implies violence against environment. In softer terms, it may be termed as organized violence against environment. Industrialization is a prominent mode of development which has to a large extent haunted the environmental frontiers. Industrialization has demanded or threatened the complete mobilization or destruction of societies (Wallerstein, 1974; Nayyar, 2006). Organized violence remains a defining feature of the contemporary global conditions (McGrew, 2007). However, in comparison with previous epochs, particularly that of the earlier twentieth century, it has acquired a substantially different form and scale (Ferguson and Mansbach, 2004). Earlier, there was organized resistance against violence but now when this violence against environment has become 'organized' the solution seems to be 'not available'. Deep Ecological Perspective also seems to be larger at this point of time and a stronger device needs to be constructed to combat the present environmental crisis. Globalization is interpreted as an economic process primarily; increasing cross-boarder interdependence for production and marketeering. "Hence, for the common motive of economic growth, the whole 'Globe' becomes one team marching ahead to exploit the environmental units and ecological richness to extract money and capital" (Callinicos, 2007). Antonio Gramsci, a popular theorist gave a theory of 'Hegemony'; at the times of being coined and developed, the concept related to the 'then two' prominent classes, i.e. 'Richer and Poor Class'. But in the present world, there exists a Global Hegemony, by the way of which the masses are ideologically convinced in context of the loss of natural wealth. Ikenberry (2007) in a paper on 'American Hegemony' quotes Ferguson who talks about the leading state, "enhances own security and prosperity by providing the rest of the world with generally beneficially public goods; not only economic freedom but also the institutions necessary for markets to flourish". Seen in this light, the hegemonic construction of a system of market openness is part of a larger process in which power, order and rules are seen (cited from Ikenberry, 2001). Then added by Ikenberry (2007), American Hegemony facilitated economic globalization in a variety of ways. At one level, the United States provided security and stability across much of the non-communist world. At another level, American Hegemony mattered as it was manifest in the layer cake of multilateral institutions and functional regimes that were instructed in the 1940s and after. These institutions provided vehicles for the provision of public goods. The theorists have been formulating theories on American Hegemony; but more generally this theory can be applied on the present global scenario. Environmental units are being excessively exploited in the name of development. In turn, trade is exploited and its virtues extolled by corporate and multinational interests, whereas environmental objectives have specifically been preserved by non-profit organization, which are generally wary of, if not hostile to, these interest. The distrust that has reflected these different traditions has led some environmentalists to forward extraordinary assertions to hostility towards globalization, and towards International Trade and the World Trade Organization and its predecessor, the GATT (General Agreement on Tariffs and Trade). GATT was attached just over a decade ago as 'GATT Zilla', reminiscent of the Japanese movie monster 'Godzilla'. GATT Zilla was considered to be a monster but this was also a time when the United States was in the throes of a national psychosis about the rise of Japan, which was widely demonized as a wicked trader, a predatory exporter, and an exclusionary importer; indeed a two faced rival. If GATT was framed and denounced, free trade itself was held to be a malign process, causing all manner of harm in all sort of ways, certainly to the environment, which was rapidly becoming a key concern (Bhagwati, 2006). As per the views given by Professor John Adams at University College, London, it is all down to how each individual view risk, in particular how we view 'nature' as a risk or the 'risk' to the nature. "Do we believe nature is benign and able to take whatever we throw at it or do we think of it as malevolent, having the power to react harshly to our interferences (Maslin, 2007). Here practically lies the base of different ecological perspectives. For instance, Global Warming Hypothesis has also been interpreted in multiple ways. Adams (1915 cited from Maslin, 2007) suggests that there are four Myths related to the 'nature' which are shown in the following figure:

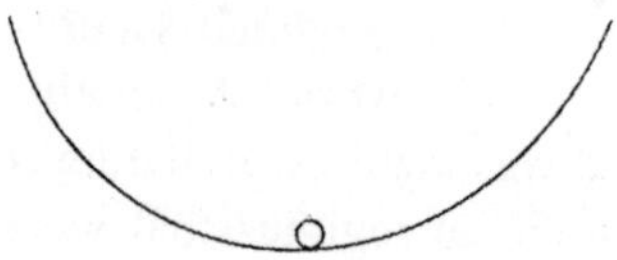

Nature Benign

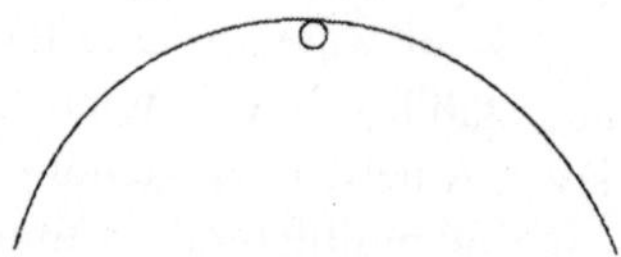

Nature Ephemeral

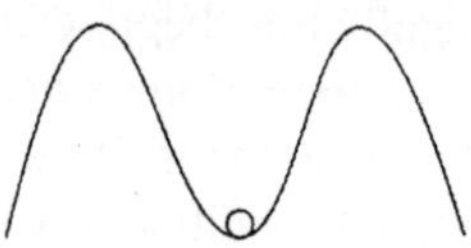

Nature perverse/tolerant

Nature Capricious

Source: See Maslin, M. *Global Warming*, Oxford University Press, New York, 2007.

Adams explains that according to the myth of 'Nature benign' nature is predictable, bountiful, robust, stable, and forgiving of any insults that human kind might inflict upon it. The myth of 'Nature Ephemeral', the nature is fragile, precarious, and unforgiving. It is in danger of catastrophic collapse thanks to human interference. The objective of environmental management must therefore be to protect nature from humans. The combination of these two myths is found in the third myths, i.e. 'Nature Perverse/tolerant'. Within limits, nature can be relied upon to behave predictably. It is forgiving of modest shocks to the system but care must be taken not to knock the ball out of the cup. Regulations are needed to prevent major excesses while learning the system to look after itself in minor matters. In this model of theoretical interpretation, the management style is interventionist. The last myth is 'Nature Capricious', which interprets Nature as unpredictable. The appropriate management style is *laissez faire*, as there is no point to management. The beliefs of this myth is agnostic concerning nature as the future may turn out to be good or bad, but in any event, it is beyond any human control. Choosing anyone of the above stated Four (4) myths to understand 'nature' and the losses occurring to it, is a multi-dimensional process. An individual has a subjective authority over choice of the myth, he intends to choose for orientations related to nature and any natural disaster or degradation these individual views are based upon many factors like one's own personal agenda including his/her survival background (rural or urban), economic status, social standards, political affiliation, etc. For instance, deforestation may be a nominal issue for some individuals, but it can be a catastrophe for others. This difference depends upon the relative dependence of individuals on the forests as a natural resource. Corresponding to this interpretation, the reaction of the masses depends. Further, in between the interpretations and reaction process falls, the 'formulation of theoretical perspectives'. In a similar way, perspectives for 'Globalization' developed. One sentiment shared by these diverse social forces is their conviction that they must protect themselves and others from the negative consequences of globalization. In this regard, they are all 'protectionists'. This term implies tendency to protect and save the ordinary masses from the adversities being forwarded by globalization. This category has been further split into 'Particularistic Protectionism' and 'Universalistic Protectionism'. Universalistic Protectionism is generalized in motive and in the approach also, thereby to establish a more equitable relationship between the global North and South. They accuse globalist elites of pushing policies that are leading to greater global inequality, high levels of unemployment, environmental degradation, and the devise of social welfare. Calling for a 'globalization from below' favouring the marginalized and the poor, they seek to protect ordinary people all over the world from a neoliberal 'globalization from above'. On the other hand the particularist protectionists think on individual grounds and blame globalization for incurring losses to individuals and their lives too. Threatened by slow erosion of old social pattern, particularist protectionists denounce globalization. They are more concerned with the well-being of their own citizens than with the construction of a more equitable international order based on global solidarity (Steger, 2007). These perspectives also formed an evidence for environmental degradation; and labelled it as one of the adverse consequences of globalization. It has never been refuted that globalization has affected the global environmental frontiers; but still, perspectives produced for the very context, further attach an ideological and social strength to a view. Environmentalists have long thought of economic globalization as a threat to environment. "Trade advocated by economists and in consequence encouraged by the bilateral and multilateral aid and development agencies and expanded by reductions of trade barriers by both unilateral actions and reciprocal bargaining by policy-makers, is a frequent target of their anguish and anger" (Bhagwati, 2006). Hence, it is clear that conflict is inevitable here and it cannot be checked to recur. Bhagwati also adds that the economists generally belong to the philosophical tradition that sees nature as a handmaiden to mankind. This humanity centric view of nature is deeply rooted in the tradition that originated among the Hebrews and Christians and spread to the western world. Spread of this ideological tradition did not stop here, rather diffused in the whole world. Hence ecological theoretical perspective, perspectives on globalization; and theories related to economic development, in an encapsulation render novel dimensions to the understanding and interpretation of ecology. But these are merely not ideological tools used for the purpose, rather in later stages, these theoretical orientation become weapon of social construction and dismantling. The theoretical perspectives related with ecology or ecological understanding provides potential for analysis and helps the mankind to balance the internalities and the externalities linked with the economy and environment.

REFERENCES

Aggarwal. S.K. and P.S. Dubey, *Environmental Controversies*, APH Publishing Corporations, New Delhi, 2002 (pp. 293-305).

Beneria, Lourdes and Roldan Martha, *The Cross Roads of Class and Genders*, The University of Chicago Press, London, 1997.

Berman, Tzeporah, *Towards an Integrative Ecofeminist Praxis*, Canadian Women Studies, Canada, 1993.

Bhagwati, Jagdish, *In Defense of Globalization*, Oxford Publication, New Delhi, 2006.

Boserup, E., *Women's Role in Economic Development*, Gower Publishing Company Ltd., Great Britain, 1970.

Callinicos, Alex, "Globalization Empiricism and the Capitalist World System", in *Globalization Theory* by Held David and Mahoney Anthony McGraw, Polity Press, Malden United States, 2007.

Capra Fritjof, "Bootstrap Physics: A Conversation with Geoffrey Chew's", in *A Passion for Physics: Essays in Honour of Geoffrey Chew*, Finkelstein and Chung-I Tang, Philadelphia, 1985.

Cranney, Brenda, *Livelihood and Environment—Local Environment and Lived Experience—The Mountain Women of Himachal Pradesh*, Sage Publications, New Delhi, 2001.

Devall, Bill, "The Deep Ecological Movement", *in Ecology*, by Clayton Merchant, Rawat Publication, New Delhi, 1996 (pp. 125-140).

Eckersley, Robyn, *Market, the State and the Environment: Towards Integration*, Macmillon Press, Melbourn, 1995.

Einstein, Albert, *Philosopher – Scientist*, Library of Living Philosopher, Evanston, 1949.

Ferguson, N., *The Rise and Fall of the American Empire*, Allen Lane, London, 2004.

Ghazala, Shahabbudin, "Forest Fragmentation, that's what is causing Bio-diversity Losses", *Down to Earth*, 15th January, 2001, Vol. 9, No. 16, 2001 (pp. 24-25).

Guha, Ramachandra, "Radical Environmentalism: A Third-World Critique", in *Ecology* by Clayton Merchant, Rawat Publication, New Delhi, 1996.

Heisenberg, Werner, *Der Teil Und Ganze*, Harper and Row, New York, 1970.

Heisenberg, Werner, *Physics and Philosophy*, Harper and Row, New York, 1971.

Ikenberry, G. John, *After Victory: Institution, Strategic Restraint, and the Rebuilding of Order after Major Wars*, Princeton University Press, Princeton, 2001.

Ikenberry, G. John, "Globalization as American Hegemony" in *Globalization Theory: Approaches and Controversies* by David and Anthony McGraw, Polity Press, Malden, United States, 2007.

Kelkar, G. and Nathan D., "Gender and Tribe" in *Kali for Women*, New Delhi, 1991.

Kuhn, Thomas, *The Structure of Scientific Revolution*, Chicago Press, Chicago, 1970.

Maslin, Mark, *Global Warming; A Very Short Introduction*, Oxford University Press, New York, 2007.

McGraw, Anthony, "Organized Violence in the Making (and Remarking) of Globalization", in *Globalization Theory: Approaches and Controversies* by David and Anthony McGraw, Polity Press, Malden, United States, 2007.

Naess, Arne, *Ecology, Community and Life Style*, Cambridge University Press, Cambridge, 2006.

Nayyar, Deepak, *Growing Globalization: Issues and Institutions,* Oxford University Press, New Delhi, 2006.

Orustein, Robert, *The Mind Field*, Octagon Press, New York, 1976.

Pepper, David, *Modern Environmentalism*: *An Introduction*, Routledge Publishers, New York, 1996.

Pojman, Louis P., *Global Environmental Ethics*, May Field Publishing Company, London, 2000 (pp. 176-186).

Robertson, R., *Globalization: Social Theory and Global Culture*, Sage Publishers, London, 1992.

Rogers, Barbare, *The Domestication of Woman*, Tavistock Press, London, 1981.

Rubin, G., "Traffic in Women," in *Toward an Anthropology of Women*, Monthly Review Press, New York, 1975.

Saxena, H.M., *Environmental Studies*, Rawat Publishers, New Delhi, 2006.

Stegar, B. Manfred, *Globalization: A Very Short Introduction,* Oxford University Press, New York, 2007.

Venkateshwaran, Reddy, *Environment Development and the Gender Gap*. Sage Publication, London, 1995, (pp. 54-80).

Wallerstein, I., *The Modern World System*, Academic Press, New York, 1974.

2

Impact of Development and Population on Environment

M.S. GILL

Population in the developing countries is increasing at such a pace that States have to adopt policies to enhance the production of food grains to meet the increasing demands of the people. In addition, the demand for consumer goods too has increased with greater stress on available resources. In the process, the resources get depleted or degraded. In fact, the process of consumption has accelerated the export and import of commodities among countries but who is going to stand at the advantageous position, depends upon population pressure and available stock of resources. The developing countries despite their best efforts to sustain resources cannot afford to activate their resource use and at the same time check depletion and degradation of environment. Human activities are increasingly depleting the available resources and degrading the environment as well. The problem had assumed such a monsteral dimension that efforts are afoot to revitalize the degraded environment through the active intervention of international bodies such as United Nation Environment Agency, United Nation Development Programme and the like. Environment has also been taken up as a human right issue and many nations have incorporated this as a Fundamental Right as in India. In this paper an attempt has been made to understand the effect of development and that of the population on environment. It equally endeavours to look into the environment awareness perspective and the role of technology in its sustainability.

Development in the Capitalist and Marxist system is based on the theory of growth, material progress and a manipulative view of techno-managerial approach to human welfare. The basic ingredients of development are materialism and technology; measured by Gross National Product of a country. Gradually the experience of development with growth has led to the incorporation of social indicators such as education, health, sanitation and so on. The third world countries added components like "self help", "basic needs", "minimum need programme" to bridge the widening inequalities among the people and nation (Sharma,1986). But development on the whole was and is still identified with physical and economic growth in terms of production, industrial and agricultural growth rates, export of commodities, commissioning of power plants and irrigation projects etc. In the modern conception, development incorporate multiple angles of life and all chores of life that are inter-knit and inter-linked to such an extent that development cannot be achieved until and unless a well regulated and an efficaciously preserved environment is facilitated and ecological issues are given due importance. Not only environment and its sustainability, gender equality and identity recognition too have been incorporated to the concept of development. The development path treaded so far has severely impaired the protective system by increasing emission of greenhouse gases, toxic pollutants and others and supportive system through erosion of bio-diversity, deforestation, soil, water and air pollution. At the same time, the principle of "growthmanship" has overstrained the natural resource base without relating it to people and nature. Such growth perspective has differentiated communities according to regions, class, sex, age etc. world experienced the first shock of development and its effect when oil prices hiked and struggle among nations to control oil reserve led to cold war between developed and oil rich nations. This has more effect on developing economies like India which was building their industries on commercial energies such as petroleum, electricity and nuclear power (Kaushik, 1996). Subsequent arm struggles led to oil slicks, burning of fossil fuel reserves in Iraq, Kuwait and the like which led to a number of environmental problems such as air pollution, destruction of aquatic life, and pollution of water resources. It is this realization that has led to the emergence of 'environmentalism' which means ideologies and practices which inform and flow from a concern with environment (Pepper, 1986).

Again, the capitalist development ushered by industrialization is considered as the prime causative factor that has been altered the man-nature relationship. The generation of surplus and non-local uses, the crux of industrialization, has eroded the traditional values pertaining to nature which have developed out of close symbiotic interaction of man with nature. These values used are justifying nature as supreme that provides protective cover and sustenance to man who is subordinate and dependent. The process of industrialization has altered the whole system and nature now has come to be at mercy of the former. This has threatened the very survival of the vast section of people, numerous natural species and even the planetary life. According to World Watch, State of the World, 2000, the use of earth resources has threatened global environment by raising temperature due to continued use of uncleaned technologies and increasing population pressure on groundwater, cropland, oceans and forests. These are speedily leading to ecological collapse. Bonan holds that human use of land especially clearing of forests for agriculture is an important cause of regional climatic changes (cited from Down to Earth, 2001). The intensive and extensive use of natural resources with materialistic approach to life and nature – central to the present model of development has adverse effect on the poor, the tribal, the forest dwellers and the women who heavily depend upon them as primary source of their livelihood. It has equally deprived these communities of their age old habitat and vital subsistence resources. Also, the long run sustainability of the resources is endangered (Kothari, 1996). Development rather than relating itself to the people and communities has taken away the resources from such communities which sustain on such resources for the benefits of consumers living in urban areas or if used by the effected communities, they have to pay a price of its processing in the industrial unit located away from their inhabited areas. In this way, the development process differentiates communities and regions.

Further, the establishments of dams and hydroelectric projects, roads and railways, industries and such other projects have environmental consequences as well. For instance, thermal power plants emanate approximately 500 metric tons of ash everyday and it is usually dumped in ponds, marshy lands and agricultural fields thereby affecting the people surviving on these resources. The super thermal plant at Roopnagar is providing slow poison because of pollution generated from it and other negative consequences; the seepage of slurry from the dykes is creating a water logging problem to a large area (*The Tribune,* 2000). According to UNDP, a large proportion of 20-30 million migrants moving to urban areas each year have a big proportion of environmental refugees whose land is so eroded or exhausted that it can no longer sustain them. The pattern of development in India has degraded more than 75 per cent of agricultural land because of soil erosion, water logging and salination and six billion of top soil is lost every year as a result of deforestation, overgrazing etc. (cited from Sheth, 1997). Similarly, the use of chemical fertilizers, toxic as they are, kill the earthworms and other bacteria that convert material on earth to natural nitrogenous and phosphoric manure and thus, prevent earth from regaining fertility in a natural way (Nath, 1996). So as to keep the pace of production, consistent application of agro-chemicals becomes the necessity of the cultivators. This shows the eco-material available to man is being used in the name of development and it leads to soil, water, and air pollution besides increasing the size of environmental refugees who move to urban areas for greener pastures and end up with increased pressure on the already limited resources of urban areas.

Equally the provision of infrastructural inputs may be useful from the viewpoint of a section of the society who wants to extract or exercise control over the resources. But it may not be able to enhance the ability or power of the people for equitable share in the resources. Development in the broader sense means enhancing the power and ability of the people to direct their lives in the context of their environment, history and aspiration for the future (Sheth, 1997). Eckersley (1995) has added social justice as a dimension to human emancipation which changes the thinking and consciousness of the people. Now with added dimension of identity, development involves enlarged range and quality of choices of lifestyles, of occupations encompassing better health, education and freedom from oppression and poverty without jeopardizing their cultural identities. This will be possible only if development process is eco-friendly and all sectors and strata of society have access to the use of natural resources in an equitable and sustainable way. The stringent rules of the Government quiet often do not allow the people to meet their aspirations of sustaining their resources. For example, the people of the drought hit areas especially of Rajasthan, Gujarat, and Andhra Pradesh have adopted a novel method of rain harvesting through check dams to meet their water needs. In the case of Orissa, check dam at Goudpali village in Nantideul of Sambalpur built by the people was not allowed completion by the Government (Down to Earth, 2001). On the other hand, the Rajasthan Government believes that it has a right over every raindrop and makes a move to demolish water harvesting structure set up by local people (Down to Earth, 2001). Looking at these, one wonders why the notion of equitable and sustainable way of access to natural resources is incorporated in the concept of development.

Since production is no longer restricted to providing barest means of subsistence for all classes, all forms of production lead to the division of the population into different classes with different, often antagonistic interests. The interest of the ruling class becomes the driving factor of production with the aim to meet the production and consumption demands of the dominant class and subsistence of the dominated sufficient to ensure surpluses. They pay little attention to natural or social effect of the surpluses. Their concern is with profit and generation of surplus, even their concern for the commodity and its purchasers is guided by their limited interest. For instance, the drugs found to be beneficial to patients suffering from tropical diseases are produced keeping profitability into consideration not the use or the population suffering from it. Only 1 per cent chemical compounds developed between 1975 and 1982 was for tropical diseases (Devine, 2005). Likewise, forests in Cuba were burnt down to obtain sufficient fertilizers for getting high profitable coffee trees. Subsequent tropical rainfall denuded the mountains leaving the people with insufficient subsistence resources. Generating surplus and profits is good enough for the development of nation's economy but depriving people of their subsistence is an issue of human survival and rights. In the same vain, a large irrigation project in Madhya Pradesh which was originally expected to increase crop production by at least tenfold has led to extensive water logging thereby reducing corn and wheat yields. According to Swaminathan (1981, cited from CSE, 1996), within 10 years of introduction of canal irrigation in Chambal area of Rajasthan and Madhya Pradesh, about 25 per cent of the command area became waterlogged. The effect of waterlogging and salinization are particularly severe in highly water-retentive black cotton soils of Maharashtra, Madhya Pradesh, Karnataka, Andhra Pradesh and in delta coastal areas (George, 1997); thereby effecting land productivity and this enforced the farmers to launch 'Mitti Bachao Abhiyan' in 1977. So, irrigation projects its raise productivity of the farm end up transforming fertile fields into 'wet deserts'. Likewise in Rayagada district of Orissa—home of Paroja Kondha tribe—the private mining companies have got mining right to extract bauxite which would eventually dry up around 700 streams and chew up forests, the lifeline of the district's 85 per cent tribal population (Down to Earth, 2001). The United States responsible for 24 per cent of the global Green House Gases emission refuses to take domestic measures till developing countries also do the same. Such examples explain the neglect of the interest of the poor disadvantageous groups or states besides infringing their right to live and sustain. At the same time it reflects the tendency to capital venture for market gains at the expense of suffering humanity.

Also the main cause of environmental destruction in the world is the demand for natural resources generated for the consumption of the rich; whether that are rich nations or rich individuals or groups within the nations. The Prime Minister of Malaysia has started at the Earth Summit 1992 that the high per capita consumption in developed countries place a far heavier burden on the environment than that in the developing countries. In this regard, per capita the US residents consume 33 times more aluminium, 183 times more natural gases and 385 times more pulpwood than the Indians. Likewise the rich are spending much higher on electrical equipment, motor vehicles and scooters (Sheth, 1997). Even among the developed nations, there is a difference in the environmental pollution caused by their populace. The Australian Institute analysis reports that average Australian Green House Gases (GHG) emissions are 30 per cent higher than those of the US citizens. An Australian is responsible for 27.6 tons of GHG emitted against one ton by an Indian (Down to Earth, 2001). Not only this, the rising temperature due to increasing emission of GHG reduces grain yield by 10 per cent (Brown, 2005). Recent research findings indicate that if temperature rises to the lower end of the range projected by the Inter-governmental Panel on Climate Change, grain harvest in tropical regions could be reduced by an average of 5 per cent by 2020 and 11 per cent by 2050. This means less availability of grains per unit, leading to undernourishment/ malnutrition. To improve grain productivity, Third World countries where population growth is high, are spending more on agro-chemicals, use ground water/other water resources, meeting subsidies and the like. With the result, vegetables, milk, food and blood samples show high pesticides residues. According to a study by Post-Graduate Institute of Medical Education and Research, there is a link between pesticide use and rise in cancer cases in the cotton belt of Punjab (Down to Earth, 2007). The study further finds varying degree of genetic mutation in the 65 per cent of blood samples. This way population pressure on demand for cereals and other agricultural products is leading to overuse of resources and rise in the price index of commodities. As a result, poor people pay more for the same thing such as food, water, drugs and the like with their little or meagre economic resources. Millions of the poor already spend 70 per cent of their income on food and rising prices will impoverish more people over period of time into poverty. When poor spend more on food, drugs etc., it is called poverty penalty, a sort of human right infringement as these people will be devoid of basic needs and right to a dignified living. Also, small stakeholders and large section of population living below poverty line (Table 2.1) especially in the developing and under-developed countries have to compete with the rich for their share of resources which they produce through their labour. The wedge between rich and poor may be one of the reasons for the rise in crime rates in such countries. Terrorism and gender inequality to a large extent are the product of uneven availability and distribution of resources.

Table 2.1: Percentage of Population below Poverty Line (India)

Year	*Rural*	*Urban*	*Total*
1983 (38th round)	45.65	40.79	44.48
1990-1991(46th round)	35.04	35.29	35.11
1999-2000(55th round)	27.09	23.62	26.10
2006-2007 (Estimated)	21.07	15.06	19.34

Source: India Yearbook 2003, Manpower Profile.

Development has further precipitated the degradation of land and forests and overuse of water, energy and other resources besides affecting the crop yield. Factories and farms to have more production are overusing at least 40 per cent more of water thereby putting a pressure on the available water. The extensive irrigation projects to tap water resource for meeting the consumption need of the factories and farms have created problem of floods, recharge of ground aquifers in addition to changing the courses of rivers and streams. In the world demand for water has climbed, water table has fallen in countries like China, India and the United States, which produce together nearly half of the world grain (Brown, 2003). People over extract and deplete available water through deeper and deeper tubewells to grow water-guzzling crops like sugar cane, rice, wheat, or even flowers (Down to Earth, 2007). There is a changed cropping pattern with the upcoming of Hirakud Dam in Orissa. Farmers are growing rice and its percentage area has increased from 70 to 98 during kharif and 33 to 60 per cent during rabi season whereas area under other crops has decreased from 30 to 2 and 15 to 10 per cent during respective seasons (Report of High-Level Technical Committee, Orissa, August, 2007). With the adoption of wheat-rice rotation under green revolution areas has encouraged canal irrigation and excessive use of ground water for irrigation. Punjab which produces 60 per cent of wheat and 40 per cent of rice of the Central Pool finds water table falling at a rate of 25 to 35 cm a year. The State of World Report, 1998 published by World Watch Institute, says the gap between water use and sustainable yield of the aquifer is so high that the aquifer under Punjab could be depleted by the year 2025 (Mahapatra, 1999). Like Punjab, Northern China is literally drying out due to falling water tables and accelerated use of water. In addition, the use of canal irrigation without proper arrangement of drainage is resulting in soils become alkaline or saline (Nadkarni, 1987). Cultivators of Abohar, Fazilka, Ferozpur and adjoining areas acutely feel these problems. At the same time the problem of water logging in these canal irrigation areas is an example of misuse of precious water resource. Farmers are now facing two new challenges; rising temperatures and falling water tables which is going to effect the productivity of soil and ultimately the per person availability of foodgrains. Of course, the poor and the downtrodden are going to suffer due to malnutrition. Shortage of foodgrains may destabilize the economy as India has to import wheat in 2006 and 2007 to meet the demand of the rising population. It is an indicator of strain on economy of the country. In order to meet such exigencies, taxes are levied and it ultimately affects the poor the most.

Further, with depletion of resources, changing course of rivers, submission of forests, and falling water table the problems of poor, the tribals, the women and the children have increased. With deforestation, communities surviving on this source find difficulty in gleaning fuel wood, fodder for cattles, and water for consumption and irrigation. On the other hand, the exploiters of natural resources continue to dominate and State supports the dominant ones to earn revenue and profits. With respect to the newly carved out States of Jharkhand, Uttaranchal Pradesh and Chhattisgarh in India, Aggarwal (2001) holds that Government of India loves to talk about poverty but usually serve the interest of the contractors and companies which exploit and destroy rich natural resource base of the poor i.e. forests and bio-diversity in particular. Even the changed crop rotation pattern to achieve high productivity in cereal crops in Punjab to feed increasing masses, has excluded oil seeds, pulses, vegetables and the like, thereby the poor masses are deprived of their cheap vegetable protein (Bhullar, 1990). Equally in the name of national interest, the poor people are asked to support the big projects such as Sardar Sarover, Cogentrix, Tehri Dam, Hirakud, and others which in reality serve the purpose of the few well-off resourced agriculturists, industrialists, and the like; depriving at the same time others of the traditional access to land, forests and water. Nevertheless, in the long run all are going to face the problem of ecological imbalance created in the name of development. So, development as well as ecological destruction disproportionately affects the poor and the rich and in this regard, environment degradation and social justice are opposite to each other.

Now going through the way urbanization is taking place in developing countries like India, the population shift is taking place more than the natural growth of urban areas. In India, the per square km population density has increased from 77 in 1901 to 324 in 2001 (Table 2.2). Among the different States, the population density is highest in Delhi (9294 per sq km) followed by West Bengal (904 persons/sq. km.) and Bihar (888 persons/sq km). This increasing inflow of population from rural areas put pressure on the available infrastructure besides generating a large mass of municipal waste, emergence of slums, unemployment, competition and conflict. With the introduction of advance electronic technology, managing e-waste is also becoming a problem of the urban areas. All this waste is dumped in the open space or landfills or water bodies thereby choking the already polluted land and water mass. The sprawling industries coming up in already congested areas are creating health problems. People living in the outskirt of industrial complexes constantly complain of headache, dizziness, rashes and chest pain (Down to Earth, 2007). A lot of industrial and municipal waste goes to drains leading to high levels of mercury, cadmium, chromium, copper and lead in ground water. In this way the very population that enjoy the benefits of development in term of profits, surpluses and income rise, face its consequence in terms of constantly accelerating cost on health. Equally a large proportion of the population is illiterate in the developing countries and they fail to understand the basic scientific issues like greenhouse effect, nutrition, hygiene, role of forests and the like. For this, their resources are exploited by the interest groups who continue their spree to degrade their sustenance resources. It is not the poor and the tribals, the rich too are not aware of the environmental and ecological consequences. Even people lack consciousness of the environmental and population changes.

Table 2.2: Population Density

Year	*Birth Rate /1000*	*Death Rate/1000*	*Density (Population /sq km)*
1901	49.2	42.6	77
1911	48.1	47.2	82
1921	46.4	36.3	81
1931	45.2	31.2	90
1941	39.9	27.4	103
1951	41.7	22.8	117
1961	41.2	19.0	142
1971	36.9	14.9	177
1981	33.9	12.5	216
1991	29.5	9.8	274
2001	25.4	8.4	324

Source: Yearbook 2003, Manpower Profile.

The question that arises: is there disjuncture between perception regarding availability of natural resources and comprehension of ecological damage? The struggle of traditional artisan fisher people against mechanized fishing vessels, local people's movement against hydro-electrical dam in Kerala and other parts of Inida, Chipko movement against felling of trees and so on were initially directed towards their very survival as their livelihood was threatened. The ecological dimension was not immediately perceived, was added later on as the mechanized fishing will destroy fish breeding grounds; destroy rich bio-diversity of the catchments areas; threaten the fragile ecosystem and so on. The ecological consciousness is also observed among the owners of electro-plating units operating in Ludhiana city of Punjab (India).The owners are adopting housekeeping method to control pollution. In their view, it controls 80 per cent of water pollution and helps reuse of collected nickel and different acids (*The Tribune,* 2001). This consciousness arises out of becoming aware of their polluted ground water and dangerous effect of the chemicals on the health of the people and the workers. Examples like this indicate that ecological awareness is limited to immediate environment of the community and governed by the relation with natural resources within it. Further, the poor have little ability to desist from using the subsistence natural produce—fuel, fodder, fiber etc.—rather wisely or with restrain.

Under such circumstances, the natural resources get neglected both in the hands of the rich and the poor. In the process, the degradation may reach an irreversible stage where the crisis of the economy and the society at large may emerge. According to Shiva (1991 : 342), "the neglect of the role of natural resources in ecological processes and its people's sustenance economy, the division and destruction of these resources for the community production and capital accumulation are the main reasons for the ecological crisis and the crisis of survival in the Third World". This prediction can be true for the world if the process of degradation continues at the present pace.

Further, the sustainability of society requires decision on technologies taking into account environmental constraints. Technical solution to specific problems can be actualized through research but the global and long-term solution to environmental problems require concomitant support from all developed and developing societies and the different sections of these societies. Not only this, political system, economic system and social system preserve ecological base of development for sustainable development through people's participation, promoting self-reliant and subsistence technologies and promote harmonious development (Lender, 1989). So far, there are no clear criteria that mutually commensurate development and environment. Kyoto proclamation on green house gases emission is one such example. In spite of this, progress towards sustainability demands measures and comparisons of the ecological price of the commodity production in addition to social and ethical ramification. National and even international bodies usually refrain from explicit normative statements and the social and economic price involved because of lack of political judgment or their being at variance with other considerations. Brown (2003) suggests that cost of production of goods should include the cost of ecological harm that good creates. This type of resolutions may escalate the cost of goods and service and it amount to denial of such facilities to the poor. Moreover, when Kyoto proclamation has been partially enforced at international level and especially those responsible for large emission of green house gases like U.S.A., how can the objective of sustainability of resources are achieved? Controversy over large hydro-electric projects like Narmada valley project in India, use of banned agro-chemicals like endosulfan, Phorates and many others are such issues that require technological inputs with environment consideration. California is now going for high efficient toilets and urinals by 2014 and this will reduce water use from 6.1 to 4.8 and 3.7 to 1.8 L respectively (Down to Earth, 2007). But do the developing nations support the cost of adoption of advanced technological or research in this area with their fluctuating economies. At the same time, it is felt that do environmentally friendly technologies have all answer to this malice. Even if a country shift from coal energy to wind power, the demand for energy for air-conditioners, dishwashers, geysers and freezers will continue to grow unabated due to population growth and increasing demand for such goods.

In these conditions, how to direct the process of development? Also, there is a notion among Third World countries that development is enemy number one of mankind and look upon non-development as the only way out. But, the developing nations cannot do away with the process of present development despite talking of treading the path of non-development as these countries have to catch the developed nations and states. Hence, deterrent tendencies of development need to be checked (Mohanty, 1998). The development supported by science and technology has strengthened the survival of man on the earth besides improving his quality of life. So, blaming science and technology for all ills do not hold well. In reality, the growing environment awareness is itself the outcome of increased scientific knowledge. Even the success of people's struggle for the maintenance and management of their eco-system owes it to the active intervention of scientists abreast with the consequences of ecological degradation and mismanagement. Rejecting science and technology and the process of development for all ecological consequences on the earth may pose a serious problem in the context of the struggle for eco-balance. The efforts should be rather to strengthen scientific knowledge so as to evolve socially rational and ecological appropriate technologies. Such technologies instead of stripping the traditional eco-friendly knowledge base developed under natural process of man-nature interaction, should fall in with it and let local communities manage their eco-material on this blended knowledge. The success stories of harvesting rainwater to meet drought in rainfed areas with the fusion of tradition and modern scientific knowledge (Down to Earth, 2000) should be the encouraging point for the modern civil society.

Conclusion

The problem of development and population growth has ecological consequences is understandable but their role in generating inequalities and widening the gap between have and have-nots is not given adequate attention. The analysis points to the ecological consequences of the development and even the very notion of treading the present path of development with its ecological ramifications. Further, available eco-material is being recklessly used and consumed by the dominant and affluent sections of society without realizing that it is ultimately going to lead to a situation of 'no return' where the difference between rich and poor will merge and both will suffer from the ecological damage/degradation. The spate of urbanization is adding to the vagaries of the common masses who feel highly segregated from the rich in terms of use of infrastructure and resources. In addition, the demand for grains is pushing the farmers towards greater exploitation of land and water besides use of agro-chemicals which further degrade the already polluted water and soil. Excessive use of resources for meeting the demands of the present generation cannot sustain the future generation's demand. Under these conditions, the concept of development redefined in term of gender equality, identity recognition and others should sound a reality not a mere academic exercise.

Equally, the affluent societies or sections of the society should redefine their interest and limits within a broader environmental perspective. It demands altering the relationship of dominance and subordination through structural adjustments in the society. At the same time, it requires a realization on the part of the rich that another section of underprivileged and underdeveloped exist whose survival depends upon the ideology of 'sharing the resources' and 'not me but you' rather than appropriating the resources and limiting to self-interest. Such an understanding will ensure judicious handling of the eco-material on which culture and civilization has developed. This will repose the faith of the underprivileged and underdeveloped communities that scientific advancements are for their sustenance and well-being.

REFERENCES

Aggarwal, Anil "Ecological Poverty", *Down to Earth*, Vol.9, No.16, Jan. 16, 2001.

Bhullar, N.S. "Development Strategy for the Nineties", *The Economies Times*, May 10, 1990.

Brown, L.R. *Plan B: Rescuing & Planet Under Stress and a Civilization in Trouble*, Orient Longman, Hyderabad, 2005.

Devine, Carol "Development Research from the Third Sector Perspective", Discussion Paper, Centre for Developing-Area Studies, 2005.

Eckersley, Robyn. *Market, the State and the Environment: Towards Integration*, Macmillan Press, Melbourn, 1995.

Engels, F. *Dialectics of Nature*, Progress Publishers, Moscow, 1972.

George, T.J.S., *India At 50, Facts, Figures and Analyses*, Express Publications, Chennai, 1997.

India Yearbook, 2003, *Manpower Profile*, Institute of Applied Manpower Research, 2004.

Kaushik, K.K. "Agricultural Policy in the light of New Economic Policy", in *Agricultural, Rural Development and Panchayati Raj* (ed.) V.S. Mahajan, Deep and Deep Publications, New Delhi, 1996.

Kothari, Smitu, "Whose Nation, the Displaced Victims of Development", *Economic and Political Weekly*, January 15,1996

Kumar, A. "Receding Water Level", *The Dainik Jagran*, December 26, 2000.

Gender, Warren. *Development*, Society for International Development, Rome, 1989.

Mahapatra, R. " Punjab facing Ground Water Crisis", *The Tribune*, October 28, 1999.

Mishra, S.S. "Toxic Nexus", *Down to Earth*, Vol. 16, No. 15, 2007.

Mohanty, N.B. "Environment and Development: Search for an Alternative Paradign" in *Sociology of Environment* by Pawar and Patis, Rawat Publications, Jaipur, 1998.

Nadkarni, M.V. "Agricultural Development and Ecology: An Economist View", *Indian Journal of Agricultural Economies*, Vol. 42, No. 3, 1987.

Nath, K.P. "Environmental Issues in Indian Agricultural Development", in *Agriculture, Rural Development and Panchayati Raj* (ed.) Mahajan, Deep and Deep Publications, New Delhi, 1996.

Pepper, D. *Modern Environmentalism: An Introduction*, Routledge Publishers, New York, 1986.

Sharma, S.L. *Development; Socio-Cultural Issues*, Rawat Publishers, Jaipur, 1986.

Sheth, Pravin *Environmentalism*, Rawat Publishers, Jaipur, 1997.

Shiva, V. *Ecology and Politics of Survival*, Sage Publications, New Delhi, 1991.

Swaminathan, M.S., "Irrigation and Our Agricultural Future" (1981) cited from *The First Citizens' Report*, Centre for Science and Environment, New Delhi, 1996.

———"The South", *Down to Earth*, Vol.9, No.13, November 30, 2000.

———"Tryst with Rain", *Down to Earth*, Vol.9, No. 11, October 31, 2000.

———"Who Owns the Rivers?", *Down to Earth*, Vol. 10, No. 4, July 31, 2001.

———"Action Reaction", *Down to Earth*, Vol. 10, No. 8, September 30, 2001.

———"Worst Polluters", *Down to Earth*, Vol.10, No.10, October 15, 2001.

———"Thermal Plant Pollution Hits 30 Villages", *The Tribune*, September 29, 2000.

———"Electroplating Units Pollute Ludhiana", *The Tribune*, January 29, 2001.

———"Eco-Friendly Toilets", *Down to Earth*, Vol. 16, No. 15, 2007.

3

Global Environment Risk

A Threat to Sustainability (A Transnational-Sociological Analysis)

JASLEEN KEWLANI

Introduction

Environment applies 'the study of the biological, chemical, and physical surroundings of man'. In simpler and authentic terms, it is a bag of benedictions and worth for the mankind. Human society can not progress in a sustained pattern until and unless the natural environment is preserved. Environmental degradation is the serious most hindrance in the way of attainment of the goal of 'Sustainability'. Mankind in the race of proceeding and developing on economic grounds has conveniently compromised the environmental onus; creating a non-sustained pattern of growth. When the concept of environmental degradation is being referred to, it transcends the natural geographical limits and be universalized in terms of effects and dimensions. Hence, the present environmental threat is not natural rather an international one, constituting a devastating 'Global Environmental Risk'. This risk is a means and also an 'end' in itself. 'End' of the developmental projects for economic upliftment and a 'Means' of further improvement. To explore the issues posed by global environmental risks, it is important to attain clarity about the concept of 'Global Environmental Change' and 'Global Environmental Risk'. Change is a part of dynamism of human life and ever developing world is bound to experience some changes in the environmental structures. The Environmental Change is of two types, (i) Systematic

Environmental Change, (ii) Cumulative Environmental Change. In case of the systematic environmental change, change at any locale can affect the characteristics of the environment anywhere else, or even the global system itself (Turner, *et al.,* 1994). On the other hand, the cumulative environmental change may well eclipse systematic changes in both long-term and the short-term consequences. This type of change refers to accumulation of regional and localized changes that are distributed widely throughout the world. But this environmental change acquires the shape of environmental risk if environmental entities start depleting or replenishing. For instance, if the types of trees being sown in a particular forest are altered with some other types of trees depending upon the changing commercial demands, it is an environmental change. On the contrary, if the trees of a forest land are being incessantly exploited for fulfilling consistently multiplying general and commercial needs of the communities living in proximity, it is the phenomenon of Environmental Risk. Environmental changes and their relative intensity pose distinctive challenges to the creation of an adequate knowledge base concerning both the environmental perturbations and the vulnerability of human and ecological system, to the establishment causal linkages in the structures of the risk. This environmental risk further shapes the societal responses and shakes the responsibility of the global state to respond effectively. The paper aims at discussing various dimensions of the Global Environmental risk with references to some prominent nations of the world.

Environmental Degradation—A Review

Sundry evidences for environmental deterioration are stratospheric ozone depletion; global warming; continued loss of bio-cultural diversity, etc. These traumas imposed upon the nature are not regional rather global in their implication. Amidst globalization of the issue of environmental degradation, its acute state in India cannot be ignored. India's environmental problems are gaining global significance because of the rapid and excessive speed of urbanization and the entangled lack of infrastructure. Increasing urbanization, industrialization, deforestation and development of the means of transportations as the other cousins of economic development put tremendous pressure on natural resources, and therefore, these resources get exploited in an unhesitant manner; pressing need to strike a balance between developmental planning and urgency to safeguard the environment (Ghosh, 2007). As asserted through various environmental analyses, there are sundry potential drivers of Environmental Deterioration. Those are: (1) population, (2) affluence, (3) technology, (4) poverty, (5) market failure, (6) policy and political failure, (7) the scale and rate of economic growth, (8) the nature of our economic system, (9) our culture and its values, and (10) the forces loosed upon the world by the globalization of the economy (Speth, 2004). All these forces lead to misuse and over-exploitation of the natural resources, which is a danger to preservation of the environment.

'Globalization' as a concept refers to the compression of the world and the intensification of the consciousness of the world as a whole (Robertson, 1992). When consciousness of the world gets intensified, it gives bounce to the quest for development; which encapsulates multiple angles of survival including that of the environment also. The process of development under globalization implies hyper-consumption of energy and natural resources, leading to environmental degradation. This kind of imbalanced development further results into the emergence of many social inequalities and issues. In such an adverse situation, it is only ten per cent of the world's population which is consuming the bulk of non-replenishable natural resources for satiating the avarice of the developed world at the cost of ninety per cent of the world population. This reflects a threatening phenomenon in which the present generation of the world is not only damaging nature and its environment against the interests of contemporary relatively less developed populations but also against the interests of their own future generation (Agarwal, 2007).

Environment can be managed if all the entities of environment are managed simultaneously, which refers to a planned and systematic approach for protection of land, air, water, forests, etc. All these natural units and components of environment are under a consistent threat. The race of mankind is racing ahead in achieving economic development so fast, that the effect is becoming a tyranny for the environment and it simply suppresses even the emergence of the onset of 'sustainability'. The concomitant relation between economic development and environment cannot be ruled out. In modern era, environmental decline and economic contractions cannot be delinked, especially in case of those societies which have developed deep roots in their soil, which provides life protecting medicine, diction, demons and totems to construct a religion of their own (Chauhan, 1998; Iyer, 1996; Aggarwal, 2002; Shiva, 1991; Smith, 1997; Singh, 2003; Punalikar, 2000; Brown *et al.*, 2002). No effort pertinent to economic development can be sustained eternal enough under the current system of massive destruction and wanton exploitation of scarce natural resources, including both renewable and non-renewable ones. Many economic activities have been a manifest cause of deforestation and even latent one at some instances. Development paradigm has produced disenchanting results, which were earlier unforeseen due to one-dimensional strategy. Due to this consequence, the third world nations have begun regarding 'development' as the enemy number one of mankind and have looked upon 'non-development' as the only way out. But, the developing nation cannot do away with the process of development despite talking of treading the path of non-development as these countries have to catch the pace of the developed nations and states. Hence the deterrent tendencies of development need to be checked. The process of development can be fruitful only if all the available natural and human resources are wisely and thriftily used, rigorously concerned, rapidly regenerated, thoroughly renovated, wherever possibly without upsetting the currently prevailing ecological balance within a geographical region (Gupta, 2007). General approach adopted for development is not featured with a balance maintaining content.

It emerges as an academic obligation to refer to 'Sustainable Development' in detail. The term 'Sustainable Development' was coined at the Cocoyoc Declaration on Environment and Development in 1970. It is such a form of development, which was advocated for, in many events like Stockholm Declaration (1972); Brundland Commission Report (1987) and many more. Sustainability is crucial criteria for development as the future can not be sacrificed to make the present contended. It is that Development Process, which meets the needs of the present generation, without compromising the ability of future generations to meet their own needs. To bring about such a society, the developmental strategy needs to be economically viable and environment friendly (Vyas, 1996). Evanescent or short-term benefits often turn to failures in long term. To attain eternal advantages, exploitation of resources, direction for investments, orientation of technological development and institutional changes must be consistent with the requirements of the present and future both. A symbiotic relationship between consumer human race and producer natural system needs to be recognized, which implies compatibility between economy and ecology (Khanna, 2000). This compatibility is missing in case of the developmental planning undertaken by Indian governance.

Construction of dams and industrial development foot a large part in process of deforestation and other forms of environmental degradation. Major objectives of construction of dams include providing a boost to irrigation, control of floods and replenishing the depleting ground water or aquifers. Other than these objectives; a latent objective of dams is to produce a sufficient bulk of electricity, to regulate and feed the industrial requirements for an efficient production. The aim of developing or developed societies is to raise per unit productivity, which is considered to be a great indicator of economic development. In this process of targeting economic upliftment, the environment suffers (Aggarwal, 1996). Construction of dams is directly or indirectly a cause of deforestation as development of industry (an aim of construction of dams), further aggravates the problem of deforestation.

Water as an environmental asset is also under consistent threat in India. The dimensions and intensity of crisis of water relatively vary in mountainous and plain regions. Water as a natural resource is under consistant threat. In the hilly zones, typically the upper catchment areas of river basins are characterized by relatively high rainfall and low population density, the optimum demand for water is less. The potential for use in local agriculture and requirement for human consumption is limited and conditioned by topography and human settlement patterns. On the other hand, the current and prospective need of water for agricultural, domestic and industrial use in plains is often much higher than the available local resource of water, creating a severe scarcity of water as inevitable feature of survival. As a paradox, both 'under use' as well as 'over use' of water results in survival cions. Crises of 'water' further gets aggravated due to misuse of water in the areas of ample availability, construction of dams using water for conversions into electricity, etc. Increased exploitation of water through storages has also led to large scale displacement of population from their habitations and significant adverse effect on aquatic life, in particular in rivers and their estuaries. This overexploitation of ground water threatens sustainability so immensely that it becomes difficult to repair the incurred losses. Scarcity of water directly affects demographic settlements along with, further adversely influencing the rest of the environmental units also (Vaidyanathan, 2006).

Not only forests and water are being victimized due to environmental degradation, air, flora and fauna, and temperature also face adverse impacts. Any environmental loss transcends the national boundaries and acquires a global magnitude. Global environmental challenges are closely interlinked. They cut across economic sectors and geographical regions. They cannot be addressed issue by one nation or even by a small group of nations acting alone. They are driven by powerful forces and will not yield to the modest efforts, the administration have been mounting (Speth, 2005).

Air pollution is another devastating dimension of environmental degradation, which has now become a widespread problem and every individual one way or the other, is suffering from its adverse effects. At local or regional levels, its manifestation are in the form of alteration in (i) visibility, (ii) intensity of sunshine, (iii) precipitation amount, and (iv) acid rain. Its global effects are: (i) change in natural climate by rise of temperature, melting of snow, (ii) increase in carbon content in the air, (iii) increase in particulates, (iv) holes in ozone layer, etc. These impacts of air pollution further adversely affect the human health, flora and fauna, and atmosphere in general (Saxena, 2006). Hence, analyzing in a broader and complementing perspective, all the entities of environment are being affected.

The Global Disparity

According to Hass (1990) and Hjorth (1994), there prevails a lack of co-operation between the lesser and more developed nations. In case of mediterranean, owing to a wise planning by UNEP, an equal distribution of laboratories ensured that developing countries have access to resource to support indigenous research and to deal more efficiently with regional marine pollution. Such a process geared to equity considerations helped to ensure that the pressure from the economically dominant countries could be minimized through the intervention of the UNEP's planning secretariat. This stands in a sharp contrast to the experience in climatic change, where the great majority of the analysts are from the industrialized countries and the scientific work has focused on the issues of primary concern to the developed world, where little political attention or funding has existed in the south, and where a widening gap between the capacities of scientists from the north and those from the south is increasingly apparent. Until and unless the nations ranging from minimum to maximum developmental standards join hands without any prejudices and complexes; the global challenges can not be combated. In the world the North and the South divide is not the only evidence of discrimination and differentiation. In many other zones of Geography, huge disparities are found in the context of environment. Whatever the case may be, a fact remains intact that in presence of mismanagement of the natural resources, the credibility of a nation deteriorates. That is why Amartya Sen, in one of his task of literary engineering argued that virtually everybody today believes in equality of something; equal rights before law; equal civil liberties and opportunities, and so on. Similarly all people would accept that all the inequalities are not unjust. Inequality of income is an inevitable product of any functioning market economy, though sundry questions are prevalent about the extent of income inequality. At the same time few people would accept in principal that inequalities in opportunities is tolerable when based on gender, inherited wealth, ethnicity or other accidents of birth over which individuals have no control. The view that there are limits to tolerable deprivation is fundamental to most societies and value system. Hence, the widening differences in terms of economy need to be balanced; in absence of which global vulnerability to environmental change may produce even more adverse effects on a universal level. The following part of the paper intends to forward a transnational analysis of the environmental situations of various nations of the globe.

Environmental Hazards in Netherlands, India, United States, and Kenya

The Netherlands and the United States share some of their top-ranked hazards like toxic contamination of water, air pollution and stratospheric ozone depletion. This is due to their similar levels and types of industrialization, as well as similarities in their environmental-management programmes (Bohm, 2003). Comparing India and Kenya, the two developing nations, they share two major environmental problems, those are; soil erosion and deforestation. As a finding of a study conducted by Bohm in 2003, India shares a large number of environment related problems in common with Netherlands and United States than with Kenya. This is due to the fact that India is compartively a developed country with major area of concentrated industrialization, where Kenya has much lower levels of industrialization. Economic growth through technological development helps meet many basic human needs in developing countries. It is clear, that increased availability of chemicals through importation and local manufacture has improved public health, food production, comfort, and labour-efficiency. On the other hand, tragic events (e.g. the Bhopal accident, the Chernobyl disaster) and long-term problems (e.g. chronic excessive exposures to pesticides) have raised questions about the way chemicals and other technologies should be introduced in the development process (Smith, 2003). Hence, the developmental strategies of many nations of the world are not universally accepted and endorsed.

Economic motives have very conveniently side-lined the environmental concerns. Nations to proceed and progress in economy and to multiply their 'capital', are incessantly marching ahead for technological developments as well as macro-level developmental projects; but the most 'direct' and 'manifest' consequence of this quest is the consistently intensifying environmental degradation. Science evolves as it responds to its leading challenges as they change through history. The problems of global environmental risk, along with those of equity among peoples, present the greatest collective task now facing humanity. In response, new styles of scientific activity are already under development. Traditional oppositions, such as those among natural sciences discipline and between so called 'hard' and the 'soft' Sciences are being overcome. The reductionist analytical world view is even studying these contents by even more esoteric specialities, creating a systematic, synthetic, and humanist approach. The recognition of real natural systems as complex and dynamic; entails movement to a science based on unpredictability, incomplete control, and plural legitimate perspective. Even after or encapsulation of all such optimist approaches, the use of scientific and technological development is not being made in the desired manner rather this use is considered to be a threat to survival of man and partly, it is ! Often, in one or the second aspect, economic development or scientific and technological advancement threatens environmental sustainability of the regions of occurrence. The global environmental issues are very different from the sorts of problems for which the traditional scientific problem-solving mechanisms have been successful. The contemporary environmental issues are more diversified, complex, influencing and hence, demanding deeper, technical and multi-dimensional mechanisms. But as a matter of concern, such mechanisms are either missing or not discovered as yet. The evidence of this administrative cum technical lapse is the ever- expanding global environmental risk and degradation featured with increasing scarcity of the environmental and natural units like water, fresh air, forests, etc.

Environmental Circumstances in Canada

Canada is such a nation, where the administration has set an essential priority to develop sustainable and environmentally viable and benign patterns of resource utilization and socio-economic development. In Canada, a special programme titled, "Sustainable Society Project (SSP) provides a specific context for exploring the ignored issues related to sustainability. It was in 1988, that SSP started exploring environmental themes and the related issues. The major aim was the in depth exploration of socio-economic and the environmental sustainability in Canada. The project enlisted scenario analysis to trace the path of Canada from 1990 into a sustainable future around the year 2030. The SSP assessed the relationships among values, life styles, and technological and economic development in the scenario, as well as the feasibility and socio-political implications of the scenario itself. The conserver-society concept was not the only vehicle for environmental arguments in Canada. A parallel exploration of the concept of "Ecodevelopment" was proceeding in the series of meetings sponsored by Environment Canada and the Canadian International Development Agency (CIDA) from 1975-1979. The term "Ecodevelopment", coined during the 1972 United Nations Conference on the Human Environment held in Stockholm, referred to the notion of environmentally sustainable development and launched the 'environment and development' debate that has been fostered since then by United Nations Environment Programme (UNEP). This form of development emphasizes on enhancing self-reliance basically on community level, maintaining cultural and biological diversity. Implicitly, 'ecodevelopment' was viewed as something for the third world, whereas the conserver society was for Canada and rest of the industrialized world (Robinson, 2003). Hence, Canada has always kept environmental priority on the top of its planning regime and the environmental circumstances are though not the best, but even not as worse as that the nations like India, Kenya, and Netherlands.

Mexican Environmental Scenario and the Environmental Consequences

Mexico as a geographical unit of the same environmentally haunted world, does not escape the adverse effects of developmental projects. Thousands of hectares of tropical forests have been destroyed in the last hundred years. The reason is again the unapologetic and unhesitant quest of the mankind to multiply the already available capital. Timber companies followed by ranchers and coffee producers converted biologically diverse, soil-protecting forests into barren and desiccated landscapes (O'Brien, 2003). Additional destruction of forests has accompanied the search for petroleum resources in the region and an influx of impoverished refugees from Central America and migrants from other parts of Mexico seeking farmland. Others environmental issues which, Mexico at present is facing are severe air and water pollution. Air and water pollution in Mexico has already reached a fatal level. This environmental deterioration has resulted into disabling of many dwellers of the nation by increasing respiratory and cardiac diseases, gastrointestinal problems, cancer, poisoning in lungs and other toxic exposures (Aguilar *et al.,* 1995; Ezcurra *et al.,* 1999).

The above mentioned notions efficiently illustrate how man's greed for economic development is leading to his own socio-cultural, physical and biological destruction. The 'means' used by him for development, have now turned to the means of his own destruction. Giving a theoretical touch at this instance, the phenomenon can be labelled as 'Marxian', because it thrives through a situation of 'Dialectics'. Mexico is one of the most accepted examples of immense environmental degradation and the related adversities. Northern Mexico experienced the year 1989 as the year of drought and heat stress. United States consider this drought with the predictions of global warming associated with green house effect. This could further be paired with alarming consequences like lower agricultural production leading to malnutrition and expensive food imports; adequate to increase economic burden of the nation. Environmental degradation in Mexico, like many other nations, has also forwarded variable degrees of social unrest. For instance, deforestation affects the glib survival of the indigenous people of the region such as the Lacandon and Maya (Mexican Communities). Economic differentiation further marks the scenario. Although in Mexico both rich and the poor segments are exposed to ambient air and water pollution equally; but the richer can afford air conditioning, bottled water, rural retreats, good nutrition and health care, but the poor people suffer the most (Pezzoli, 1998). Forests produce an atmosphere, temperature and even habitat for facilitating dwelling or survival of animals, birds, flora; but the clearance of forests decimates this population of species of flora and fauna, which provides the food and fibre base for traditional hunting and gathering for the people's livelihood. Further, it affects and imperils the health and nutrition of poor refugees and migrants who therein are forced to survive in deforested, infertile and unproductive deforested land, resulting into food crises. Hence as a consequence of ever occurring and diversifying environmental degradation, social unrest also stem out from the social and ecological impoverishment of the region (Liverman, 2003).

The State of India's Environment

'Development' in India was expected to bring benefits to all segments of the Indian demography, but the consequences were adverse. The notion of development along with bringing economic upliftment and self-sufficiency imposed environmental disability also. Development has emerged as a process of exploitation and violence against man and nature both. The emphatic discourse of economic growth and industrialization dominant in the 1950s lost its vigour when it became obvious that large parts of the population would not profit under the existing development strategy (Linkenbech, 2007) with a priority given to industrialization the agricultural sector had been significantly neglected. Neither land reform, nor structural adjustments and technological investments in favour of the rural population had taken place (Parekh, 1995). Firstly, by stressing more upon the industrialization; air pollution, water pollution and deforestation become prominent invitees and secondly, by facilitating the agrarian frontiers for catering to the industrial needs. All this directly result into rural life deterioration. Excess and an unbalanced bias for economic development through industrialization, development of science and technology; construction of dams for generating power for the industries; ever increasing population; expanding residential areas and cities; mushrooming of Special Economic Zone (SEZs) for regional and national economic growth, etc. have all together posed an unremarkable threat to India's environment, further creating innumerable adversities for the indigenous masses and other species surviving in the national boundaries. The environmental units like forests, air, water bodies, soil, etc., become victim of these developmental projects. India's environment is one of the most threatened environments amongst various other States of the world. India is dominantly a rural society, which largely depends upon ecosystems of survive and grow. If those ecosystems are not adequately preserved, the rural spirit may perish. The phenomenon may result into further deterioration of the already backward rural societies. Even urban societies of the nation do not escape environmental degradation and its fatal consequences. In India's three main cities like Mumbai, Kolkata and Delhi, between 40-50 per cent of the dwellers live in informal, congested, often illegal slums. Not only slums, but other areas also suffer due to lack of efficacious management of the environment. Providing all facilities to everyone is a colossal task, but at least minimum level improvements must be undertaken. Overpopulation in India is the root cause of all environmental issues perpetuating in the national geographical boundaries. Over-crowding leads to emergence of many socio-medical issues and survivalhood complications. Mumbai's immigration scene presents a bombarded influx of people per day. Here, daily garbage generation ranges more than 3-20 tons, leading to 9 to 20 per cent lungs disease incidence. Noise pollution is 72 decibels in many areas of the nation. According to Jayapalan (2002), pollution of air in slum areas is equal to smoking of 30 cigarettes per day. A specific component of India's environment is that urbanization associated with growth of industries and trade contributes efficiently and magnificently in environmental pollution and toxication of ground water and soil. Industrial chemicals released by iron and steel production units contribute to nearly 70 per cent of the effluent released; on the contrary 20 per cent of their chemicals are used for the industrial output. The unregulated disposal of pollution occurs through drains joining into rivers, streams and reservoirs. This is because the industrial estate lacks waste water treatment systems, leading to all over pollution of water and contamination of the drinking water also (U.S.A.E.P., 1996). The condition of solid waste management further provides an infinite impetus to environmental degradation to occur. As per the estimates, about 100,000 metric tons of municipal waste is generated daily in India. Per capita waste generation ranges from 200 gms. to 600 gms. per day. The collection efficiency is 70 to 90 per cent in metro cities and about 50 per cent in smaller cities. The Urban Local Bodies (ULBs) spend about Rs. 500 to 1,500 per ton for solid waste management. Out of this expenditure, 60-70 per cent is spent on street sweeping and collection, 20-30 per cent on transportation and even less than 5 per cent on final disposal (HUDCO and Habitat, 2001). Another environmental loss incurring phenomenon is increasing in number of private vehicles. People now prefer private vehicles over the public transportation modes; as a matter of status, individual convenience along with time management. Hydro carbon omission from these innumerable vehicles is very high because a large part of the air fuel mixture escapes unburnt into exhaust. So, about 20-25 per cent of the fuel used is lost this way. This incidence is of a severe concern to India where near about 75-80 per cent of registered vehicles are two wheelers and three wheelers. This is how the human race is itself running and destroying the environment and atmosphere, in order to attain economic development and technological advancement. Though these two conditions are important to raise standard of living, but the curiosity of social exposure must be balanced with that of the health and environment concerned factors.

The preceding part forwarded a 'peep through analysis' of environmental circumstances of various nations like Kenya, Canada, India, Mexico, Netherlands and United States. The information states that there is an ocular threat to the global environment; which is actually a threat to the sustainable futures of all the victim nations. The development regime and efforts needs to be mingled with an approach of 'balance' and 'anticipatory assessment'. To attain 'sustainability' various mechanism need to be framed. But, striving for a sustained future is not mere on obligation of the governments of the concerned nation. Since environmental degradations offers social, culture, economic and physical implications and affects, the solutions also need to be multi-dimensional.

Global Environmental Risk: The Needed 'Defence Mechanism'

Incorporating social visions into the discourse about the future would serve the purpose of explicitly introducing a number of ignored or neglected elements into the 'managing of the planet' discussion. Strategies for the avoidance and mitigation of risks related to technological and other hazards thus need to be blended into development strategies. In this way, perhaps the developing countries can enhance well-being without experiencing unnecessary episodes of human illness, ecosystems degradation, and threatening environmental risk (Smith, 2003). The nature and the source of the specific risk need to be analyzed for solving any environmental issue. That is to say that the criteria for evaluating and managing environmental risks and the hazards of development tend to be fragmented according to the source of the risk. The source is relative, depending upon different nations, stats, cities, villages and communities; as the survival pattern and the cultural embodiment of a society determines the forms and further, the impacts of environmental degradation. Hence, research in the field of environmental need to be intensified and increased if the 'means' of the degradation of environment are to be located.

Humanity is now witnessing a growing awareness among all those concerned with global risks that no single cultural tradition, no matter how successful in the part, can supply all the solutions for the problem of the planet. Forms of knowing other than those fostered by modern western civilization are also relevant for an explanatory problem—solving dialogue. It simply implies that all the cultures and sub-cultures, which are the integral components of the society experiencing environmental deterioration, must be incorporated in the defense mechanism being framed to combat the issue. In direct terms, involvement of cultural entities means the involvement of communities dwelling in the areas. To solve environmental issues, the most potential instrument is enrollment of common masses; Non-Governmental Organizations (NGOs); Community Based Organizations (CBOs) and the Community Social Organizations (CSOs), in the solution procedures. For widespread issues like environmental degradation, the cause root of which is the human activity only; masses need to be taught about their role and duty towards the scenario. Other than NGOs, CBOs or CSOs no other element can be as constructive and productive means of 'popular mobilization' (Kothari, 2006). Successful Institutions charged with managing global environmental risks are likely to be those that develop a management approach and culture in which learning and adaptation thrive. They are likely to be institutions with greater openness and permeability in their boundaries so that greater information sharing occurs with constituents, clients and critics (Kasperson, 2003). The locus of institutional learning and innovation will expand to include lower as well as higher levels of the organization.

Global Environmental Risk is not only about 'threat'; it is also about 'opportunity'. Where it damages the flora and fauna of the nation, there only it provides an opportunity to control the negativities getting multiplied and preserve and protect the environment. Society is to deal more adequately with the ongoing environmental changes that threaten the global well-being. It is abundantly clear that at the dawn of the Twenty-first century, the past is an imperfect guide to the future. Accumulated and growing environmental stress at the global scale is increasingly dominating the risk portfolios at national, regional, and local scales, signalling a mismatch between institutional jurisdiction and future risk challenges. The burden on society to identify and anticipate such changes in advance of the experience of their impact has grown enormously, and it is apparent that future surprises will be numerous and wrenching. Referring to a holistic approach to the environmental studies; it can be stated that social, cultural, economic, physical and even psychological efforts are required to be invested in the realm of Environmental Management without which, resolving environmental issues is simply an impossible target.

REFERENCES

Aggarwal, Anil and Sunita Narain, *The First Citizens Report 1982*, Centre for Science and Environment, New Delhi, 1996.

Aggarwal, S.K. and P.S. Dubey, *Environmental Controversies*, APH Publishers, New Delhi, 2002.

Aggawral, Arvind, "Globalization, Development and Environmental Degradation: A Human Rights Perspective", in *Human Rights, Development and Environmental Law—An Anthology* by Satish C. Shastri, Bharat Law Publications, Jaipur, 2007.

Aguilar, *et al.*, "The Basin of Mexico: Regions at Risk: Comparison's of Threatened Environment" in *Global Environmental Risk* by Jeans Kasperson, *et al.*, United Nations University Press, London, 1995, (pp. 304-366).

Ascher, Williams, *Forecasting: An Appraisal for Policy Makers and Planners*, Baltimore, John Hopkins Press, London, 1978.

Bohm, Vicki, Norberg, "International Comparisons of Environmental Hazards", in *Global Environmental Risk*, by Jeans Kasperson, *et al.*, United Nations University Press, London, 2003.

Brown, R. Lester and Orient Longman, "Ecology over Economics", in *Down To Earth*, April 30, Vol. 10, No. 23, 2002, (p. 56).

Chauhan, I.S. and Arun Chauhan, *Environmental Degradation—Socio-economic Consequences*, Rawat Publications, New Delhi, 1998.

Express News Service, "M.P. from Tiger Caucus; to meet P.M. by weekend" in *Indian Express*, New Delhi, 28 April, 2005.

Ezcurra, *et al.*, *The Basin of Mexico: Critical Environmental Issues and Sustainability*, United Nations University Press, Tokyo, 1999.

Ghosh, Sancheeta, "Concern for Environmental Degradation in India's Planning: A Review", in *Population and Enviornmental Linkages*, by C.P. Prakashan and R.B. Bhagat, Rawat Publication, Jaipur, 2007.

Gupta, S.P., "Environment, Sustainable Development and Traditions", in *Urbanization, Development and Environment* by V.S. Phadke and Swapna Banerjee Guha, Rawat Publications, New Delhi, 2007.

Hass, Peter, *Obtaining International Environmental Protection Through Epistemic Consensus;* Millennium 19, 1990, (pp. 347-364).

Hjorth, Ronnie, "Baltic Sea Environmental Co-operation; The Role of Epistemic Communities and the Politics of the Regime Change" in *Co-operation and Conflict*, No. 1, 1994 (pp. 11-31).

Housing and Urban Development Corporation (HUDCO) and U.N. Centre for Human Settlement (HABITAT), *The State of Indian Cities*, Nairobi, New Delhi, 2001.

Iyer, K.G., *Sustainable Development—Ecological and Socio-cultural Dimensions,* Kay Kay Publishers, New Delhi, 1996.

Jayapalan, N., *Urban Governance,* Atlantic Publishers and Distributors, Raj Kamal Electric Press, New Delhi, 2002.

Kasperson, E., Roger, *et al.*, "Introduction: Global Environmental Risk and Society", in *Global Environmental Risk*, by Jeans Kasperson, *et al.*, United Nations University Press, London, 2003.

Khanna, P. and Babu Ram, "Environmental Education and Economic Growth" in *Environmental Management and Audit,* by P. Rao, Sasi Bhushan and P. Rao Mohana, Deep and Deep Publications. New Delhi, 2000.

Kothari, Miloon, *Development and Social Action,* Rawat Publications, New Delhi, 2006.

Linkenbach, Antje, *Forest Future*, Permanent Black Pauls Press, New Delhi, 2007.

Liverman, M. Diana, "Vulnerability to Drought and Climatic Change in Mexico", in *Global Environmental Risk,* by Jeans Kasperson, *et al.*, United Nations University Press, London, 2003.

Liverman, M., Diana, "Vulnerability to Global Environmental Change", in *Global Environmental Risk,* by Jeans Kaspersons *et al.*, Rawat Publications, New Delhi, 2003.

Mohanty, Narayana, Rabi, "Environment and Development; Search for an Alternative Paradigm", *Sociology of Environment* by Patil and Pawar, Rawat Publications, New Delhi, 1998.

O' Brien, Karen, L., *Sacrificing the Forest: Environmental Social Struggles in Chiapas,* Boulder, Co.: Westview Press, New York, 1998.

Parekh, Bhikhu, "Jawahar Lal Nehru and the Crisis of Modernization", in *Crisis and Change in Contemporary India,* by Upendra Bakshi and Bhikhu Parikh, Sage Publishers, New Delhi, 1995.

Pezzoli, Keith, *Human Settlements and the Planning for the Ecological Sustainability—The Case of Mexican City,* MIT Press, Cambridge, 1998.

Punalikar, S.P., "Inferences between Development and Environment; A Micro-narrative on Tribal Society in Gujarat", in *Resource Management and Contours of Development* by B.C. Barik, Rawat Publications, New Delhi, 2000.

Robertson, R., *Globalization: Social Theory and Global Culture*, Sage Publishers, London, 1992.

Robinson, B., John, "Exploring a Sustainable Future for Canada", in *Global Environmental Risk*, by Jeans Kasperson, et al., United Nations University Press, London, 2003.

Saxena, H.M., *Environmental Studies*, Rawat Publications, New Delhi, 2006.

Shiva, Vandana, *Ecology and the Politics of Survival*, Sage Publishers, New Delhi, 1991.

Singh, R.B., *Human Dimensions of Sustainable Development*, Rawat Publications, New Delhi, 2002.

Smith, Keith, *Environmental Hazards*, 2nd Edition, Routledge, New York, 1996.

Smith, R. Lirk, "The Risk Transition and Developing Countries", in *Global Environmental Risk*, by Jeans Kasperson, et al., United Nations University Press, London, 2003.

Speth, Gustave, James, "*Global Environmental Challenges*, Orient Longman Private Limited, New Delhi, 2004.

Turner, B.L. and Benjam, Petricia, "Fragile Lands and their Management", in *Agriculture, Environmental Health: Towards Sustainable Development in 21st Century*, University of Minnesota Press, Verman, Minnerpolis, 1994a.

United States-Asia Environment Partnership (US-AEP), "*Environmental Infrastructure Strategies in India*, Washington D.C., 2001.

Vaidyanathan, A., *India's Water Resource—Contemporary Issues*, Oxford University Press, New Delhi, 2006.

Vyas, S. Vijay, "Environmental Policy and Policy Implementation in India" in *Environment and Development*, by M.S. Rathore, Rawat Publications, Jaipur, 1996.

4

Sustainable Society
Inhibitions and Facilitations

M.S. GILL

At the threshold of Twenty first century, there is a growing realization in the world that earth is under stress due to increasing population, over and irrational use of natural resources, and degrading environment. There has been rhetoric political and legislative efforts to save environment, reduce pollution, preserve wildlife, and create more environmental awareness. Industrial polluters, obsessed profiteers are under continuous attack but reality is that human being do not live on bread alone, he has a mind, thought and ideas which he tries to translate in his life. The concern for economic growth has made material goods more plentiful and at the same time the quality of life and its attributes have changed these in value. Pollution level follows an increasing trend with growing GDP of any nation and pollution generation is an unavoidable part of any growing economy unless checked with proper environmental management (Yedla, 2000). Moreover, the struggle to gain social, political and economic power has given way to many new theoretical models to explain the allocative choices to the conception of capitalism and its relation to political and social life. This ideological framework to support the generation of profit and surpluses has eroded the relationship between man and his environment. Technology, distribution of natural resources and economic growth has differentiated nations into developed and underdeveloped. Development is increasingly defined in terms of industrial growth and use of science and technology in the production process. No doubt, other indicators like environment, gender equality, cultural identity and the like were added later on, but the symbiotic relation between man and his environment was overtaken by the conception that man is independent of nature and has potentialities to correct the negative urges of nature. This coupled with absence of an acceptable definition of a reasonably good quality of life has given birth to tendencies as individualism, egoism, materialism which ultimately impairs the symbiotic relationship that exists between men, groups and between man and nature. The symbol of development which can raise the quality of life is taken to be the path of industrialization and economic growth. Global warming, ozone depletion, toxic pollution of land, air and water and erosion of bio-diversity are some of the consequences of such developmental ideology (Singh, *et al.*, 2005). Knowing very well that this development process has divided the globe into two disparate worlds with different needs to sustain (Aggarwal, 1992), the process of industrialization and technological growth is continuing unabated. Even the developing nations are putting their might to catch up with this tendency with little realization that they are going to exhaust their resources more than the developed nations. It has threatened the world with possibilities of conflict that humanity needs greater efforts to live/survive on the earth. Loss of human values is associated with such development. In this context, the world especially the developed nations have floated the idea of sustainable development and sustainable society.

The conception of sustainable development implies inter-generational sensibility with respect to the use of natural resources but with the skewed distribution of control and use of these resources, the concept of sustainable sounds hollow. It is known fact that developed countries comprising roughly 25 per cent of the world population consumes 75 per cent of the world resources (Sharma, 1996). Accordingly, the control and use of resources generate regional inequalities about which the concept of sustainability has nothing to do. Under such circumstances, the inter-regional disparities (which are equivalent to inter-class and inter-caste inequalities) is unable to put a curb on the use of resources for those who are venturing to attain equality with developed nations (upper class/upper castes). The inequalities generated through the control and use of resources provides nations with economic and political power which is pre-requisite of trading and power patterns in international politics. Wars have been fought to result in changes in the patterns of control of resources; even though these changes may never have been publicly declared intentions. Politicians usually prefer direct control because direct control enhances their own power. On the other hand, those who are controlled have the myth that regulations through the law are effective to discourage any action which created differentiation as an anti-social. Whenever a large number of people involve in polluting, politicians are notably reluctant to enact tough anti-pollution legislations. People continue to use existing inefficient technologies like stoves, *chullas* (traditional hearth), and even obsolete vehicles, machinery etc. More than 1/3 of the global population continue to cook and heat their homes by burning traditional biomass such as wood and dung and 1.6 billion have no access to electricity at all (Maio, 2007). For them, there is no way out to go for improved technology for cooking and heating. Likewise, politicians are reluctant to make drivers the culprit in auto pollution and traffic accidents. They even allow the slums to come up on the public property with their eye on electoral power. They do not only allow them to flourish but also recognize as part of the larger urban structures by providing amenities like electricity, ration cards, unmetalled approach road with scant respect to quality of life about which rhetoric is thrown time and again. Rarely does such arrangement promote harmony and just social order. It rather perpetuates inequalities and loop-sided development at different levels in a manner that certain groups/nations continue to be dictated by terms and conditions of the developed ones.

Similarly, the bureaucrats prefer control over the market forces in the grab of environmental legislations. They need control to budget their unmet social needs. They show their concern for the environment and want to protect environment, but restrict their choices and behaviour to lesser causes or follow the command of the politicians and allow activities to continue till some environmental damage is reported. In Punjab, the popular policy is to offer free electricity and water to keep the peasantry within its vote bank with faint recognition that it may encourage overuse of water and electricity, the deficiency of which is a perpetual phenomenon of Punjab. According to a report, the introduction of subsidized power in Punjab has increased the use of tubewell as a source of irrigation (increased from 57 per cent in 1990-91 to 71 per cent in 2006-07) more than the canal irrigation which came down over the years from 42 per cent in 1990-91 to 28 per cent in 2006-07 (Dhaliwal, 2008). Not only this, the subsequent fall of water is now being drawn by a 12hp pr pf higher power motor, leading to four times increase in power consumption (Virk, 2008). In all, the falling water table has led to increase in power consumption by 2.5 to 3 times in the agriculture sector in the State. This change has made overuse of sub-soil water as well as power but certainly, fetches sufficient votes for coming to power. In the same vain, the local transporter continue to use kerosene mixed diesel for their vehicles which not only increase pollution level in urban areas but also increase the cost of maintenance of the vehicle. When Delhi Government instructed shift from diesel to gas, many of the diesel operated polluting vehicles shifted their area of operation to Punjab and Haryana as the implementation of pollution laws are not taken seriously. There is more stress on pollution check for the private vehicles as against the public ones. Such attitude is equally observed in US. US army corps of Engineers is polluting the Michigan Lake as they have the obligation to keep the canals carrying sewage clean not the lake. So, they are not under obligation to keep the lake clean and are dumping their waste in the lake (Mitchell and Simmons, 1998). Do such actions on the part of politicians, bureaucrats and other stakeholders, amount to sustain and manage the quality of life?

Likewise, the polluters or consumers of natural resources in excess who deplete the common goods like fresh air, water, beaches and the like are not accounted for assessing the level of pollution and its cost for the future. This should be implemented for all irrespective of their level of economy or political strength. To safeguard the interest of the future for sustainability of resources, these polluters should be taxed as per their contribution to pollution or degradation of the resources. Greenpeace campaigners want auto-manufacturing industry to label their cars and disclose information about contribution of their vehicles to carbon dioxide emission to enable the consumer to make informed and responsible car buying choices (Sharma, 2008). Such practices are already in force in countries like China, Japan, Taiwan and South Korea. Even for the global issues, such taxation may restore the resources for use of the future generation. No doubt, international facility is created to address global issue of environment but such facility should not differentiate nations as developed or underdeveloped. For instance, ozone depletion which is the result of over consumption of a few is taken as a global issue and hence, international facility is provided to monitor it. On the other hand, desertification and soil erosion, again an environmental issue, remains a local issue and no international funding agency support it. Only at the recent summit at Bali, it was agreed to launch a UN fund to help the poor nations meet damage from climate change like drought and rise in sea level (Sharma, 2008). Of course, such partisan system distorts the environmental priorities and how well the future of the coming generation be secured? Hence, the concept of sustainability is not easy to attain with the given system of development which ignores certain environmental issues and work with narrow mandate rather than maintaining a higher value order.

The available course of action is to promote those fabrics of social system which strive for attaining the highest form of value system whereby equality, universality and the like may be attained. In the view of Comte (see Coser, 1971), the new positive order would have 'love' as its principle, 'order' as its basis and 'progress' as its aim. On the contrary, the positive philosophy directed towards scientific growth has eroded all these basis, aims or principle; instead greed, lust, domination and unequal power relations have degraded human relations and mind by 'thingifying' everything including human being. This eroded the natural base of society with stress on economic development with greater emphasis on controlling resources may be natural, intellectual or physical. Such tendency has not only polluted the mind but the soul besides the physical world. As science and technology have failed to deliver any lasting solution to environmental degradation and establishment of a civil and ecologically responsible society; the solution has to be found in the historical path of consciousness of mind and an approach of conceptualization of reality in the existence of a real world rather than an abstract one. Life on earth has survived as a totality of organic living systems (Yong, 2000) with conditions that have been more conducive for existence and growth.

In reality, the process of development has eroded the natural systems. The health of natural systems is directly connected to our own physical/psychic health (Dodge, 1987). In the earlier historical growth of the society/life on the earth, the natural systems were meeting the subsistence needs. In consonance with it, developed cultural behaviour regulated by customs and traditions and enforced through family, kinship or community. The rules and regulations indoctrinated by religion not only sustained the system but also strengthened these to the benefit of social life. Comte (cited from Coser, 1971) holds that *religion furnishes the unifying principle, the common ground without which individual differences would tear society apart. Religion permits men to overcome their egoistic propensities and to transcend themselves in the love of their fellow men. It is the powerful cement that binds a society together into a common cult and a common system of beliefs. Religion is at the root of social order.* But the division of labour, of course foster the development of individual capacities by creating a sense of dependence on others, keeps him away from the feeling of common interconnection and that of the whole. Such development favours progress and growth but the resultant effect is differentiation between units which consolidate individual to society. It promotes struggle between individuals and groups for progress and growth which differentiate between individuals and groups in terms of rewards and privileges; resulting in tensions and conflicts in society.

Further, the process on industrialization and economic growth has been possible with division of labour and scientific development. This has promoted secondary/contractual relations mediated among groups and individuals (Table 4.1). Such relations promote organic solidarity that develops out of differences, rather than likeness, between individuals (Durkheim in Coser, 1971). For him, elements in a differentiated society are less strongly tied to common collective routines but are much more interdependent to develop a network of solidarity. The attempt to base social living on contract between individuals (groups) was inspired by hubris or madness, or both in the view of Comte. Men have duties rather than rights. Those groups or nations which construct their society on the basis of rights end up with the destruction of natural rights of the future as well as those who are in subordinate position or lack economic resources (underdeveloped). How come that solidarity will emerge? de Bonald (quoted from Coser, 1971) states *that we are bad by nature and society perfects man in his intellectual and physical aspects.* But the quest for advancement has its root in the exploitation of natural resources for the benefit of humanity. Interestingly, the very character of the present development is transforming the nature into products, no longer for simple subsistence but for a variety non-local uses and generation of surpluses (Harvey and Hallet, 1978). This benefits the appropriating classes who marginalized the others and radically change the patterns of utilization of products. When each is pursuing his own best interest in a society, it believes in the freedom of the commons. Freedom in commons brings ruin to all. Religious institutions keep individual propensities and drives in check and supplement a moral community/society and group cohesiveness. In the view of Comte (see Coser, 1971), the society of men decays into a collection of isolated, insulated and acquisitive persons if moral sanctions that tie superiors and inferiors are not there. Thus, religious ideology essentially acts as a binding force to curtail selfish, individualistic urges and desires. Such higher values need to be promoted so as to have a society that sustains the needs of the future generation as well.

The State justifies and functions to maintain by force the condition of existence and domination of the upper classes against the dominated ones. What society gains in the process is inequalities, self-aggrendization, individualistic pursuits, materialistic form of living which promotes not solidarity but conditions of conflicts.

Table 4.1: Social Expression of Development

Process	*Strength*	*Expressed In*	*Virtues*	*Lead To*	*Outcome*
Development	Secondary/ Contractual Relations	Groups, Individuals	Egoistic, Individualistic, Self-Assertion	Conflicts/ Struggle	Ecologically Impaired Society
Sustainable Development	(i) Judicious use by the Present (ii) Needs of the Future	(a) Kinship, Family, Community	Cooperation,Accommodation, Social Security	Orderly Relation	Ecologically Sustainabe Society
		(b) Groups/ Individuals	As of Development	Conflicts/ Struggle	Ecologically Impaired Society

Not only this, man to meet his primary and secondary needs engage in antagonistic cooperation, leading to formation of antagonistic classes. These classes are structurally interrelated but the interest of the prime movers is to hold, directly or sublimated, their interest above that of those subjected to them. For the ascending classes, the existing property relation in which land is held by the peasantry is a fetter upon further industrial and economic development. Acquisition of villagers' land with the promise of job and house, in the view of the displaced community may be beneficial for the industrial growth and symbol of prosperity for the ascending class but it is nothing except exchange of poverty for greater deprivation and marginalization (Narain, 2008). The industrialist want their land, water, minerals but not their labour and establishment of industries rather than sustaining a better life to the poor whose land is so acquired leads to their exploitation, marginalization and end of livelihood security. In other words, the new social relationship results in contradictions and conflict. This conflict supported by struggle among groups to gain economic and political power leads to nepotism, corruption and denigration of system of beliefs, knowledge and medicine and so on. A simple example is that burning of medicinal plants to produce purifying smoke kills a number of germs which otherwise burning of wood cannot attain. The study by National Botanical Research Institute, Lucknow found that bacteria count decreases by 90 per cent after burning of ½ kg of *havan samagri* within a hour (cited from Down to Earth, 2008). Iranian Scientists and Asian Agri-History Foundation in Hyderabad have given ample support to this justifying the knowledge that ancient communities had developed with their interaction with nature is still relevant despite adverse comments on the use of such techniques by the modern scientific community. There are so many examples like this. The basis of the modern health services are the use of chemicals and their compounded nature which may provide immediate relief but not with its side effect on the health. Of course, the evolving relationship whether it is between groups and societies or between society and nature results in mistrust among the interacting communities or societies. This may lay the foundation of disequilibrium socially and ecologically (Fig 4.1). This is the intrinsic property of the class society (Raghunandan, 1987) that put premium on the western knowledge or the knowledge developed through modern means. This impairs and in reality has impaired the environmental sensibility of the masses and as such the nature is/will continue to be treated as a commodity which can be transformed according to the needs and demands of the dominating ones whether these are nations, communities or groups. Under such circumstances, values of higher order such as others before self, live for others (Comte), equality and the like-pillars of a sustainable society are something far off to be achieved.

Moreover, prior to industrialization, urbanization and technological growth, the community was responsible to maintain common property resources. Many of the groups were driving their sustenance from these common property resources. The loosening of the social bonds strengthened by the former (acquisition of cultivable land for industries, Government machinery for maintaining forests, cutting of forests, changing character of the forestry and other) has dislocated or alienated the individuals especially the tribal and small peasantry in rural areas who feel greatly marginalized in the process of development. The new arrangements have brought conditions of relative normlessness in the whole society more particularly among the tribals and rural one. It is a condition in which individual desires

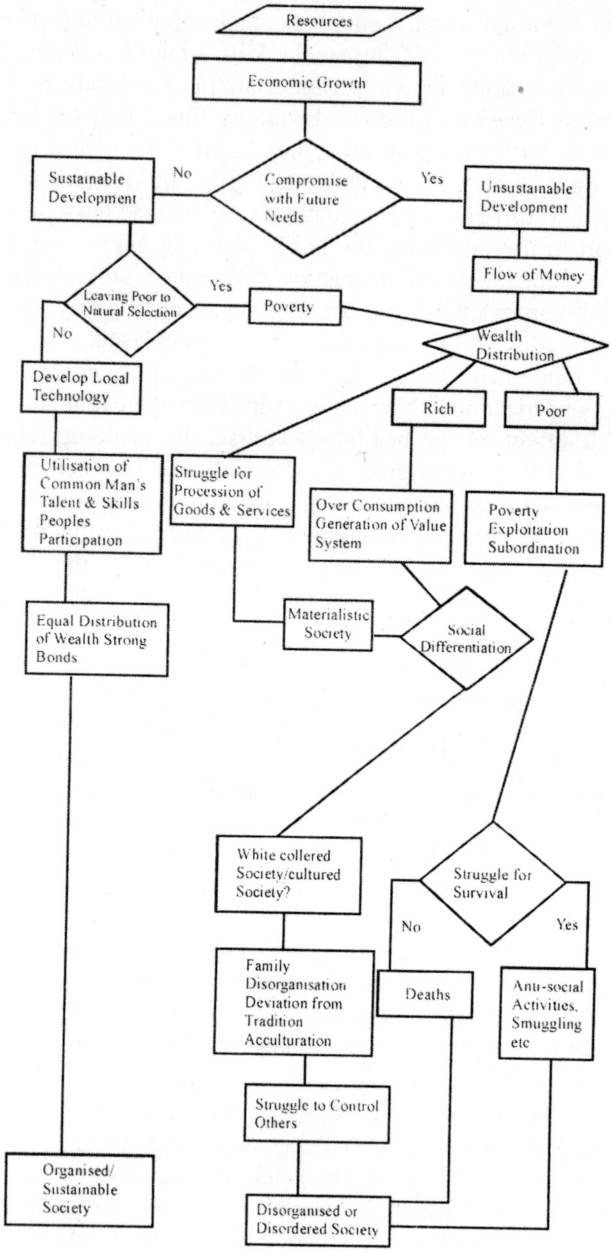

Fig. 4.1: Resource Use and Impact on Society

are no longer regulated by the common norms and individuals are left without moral guidance in the pursuit of their goals. The customary expectations from the State and the fellow men are no longer in a position to sustain the social relationships within the community or groups. These poor communities are left to sustain themselves and in the process their common talent and skill, knowledge etc. becomes obsolete in the context of larger developed community. Such deprivation may produce a one-dimensional man (Marcuse) and homeless mind (Berger). When these people compete with the others who have resources at their disposal, they are prone to lead a life in a disorganized manner devoid of all what man is trying to achieve with development (a dignified and a quality of life). To sustain in the system of society may encourage their involvement in crime, delinquency, and violence. No doubt, the uniform relations between antecedents and consequent events may not be relevant but is the worth considering causes for the foundation of a disintegrated society. Empirical generalizations and theoretical formulations sufficiently support the above.

Also, the new social order that is emerging has changed the structure and composition of the family. The joint family system which was the integral part of rural and to a great extent of the urban society is on the verge of complete collapse. Nuclear families and now single families are on the rise. The social bonds that used to integrate individuals into family, kinship and community no longer sustain. Secondly, the demand of a better social and economic life has limited the interaction within the family itself. The children are left to fend themselves and the parent's responsibility is limiting to providing material necessities, the area of which is expanding with globalization and open market. The customs and traditions that use to force regulations in the family are no longer valid. The new arrangements are more of contractual type rather than mutual inclusive ones. Not only this, there is further denigration of neighbourhood and community relations with the result that cooperative, accommodative and emotional and social security like virtues are no where to define social life. Their expression was family, kinship and community and with their changing role and structure, such virtues are not internalized by the coming generation. It is self interest and struggle for its achievement that has given way to conflict, tension and contradictions. Does under such social conditions, an ecologically sustainable society be thought of? In addition the social order that prioritized economic to social life can sustain or strengthen a value where one can think of the interest of the future generation? Moreover, more number of nuclear or single families means more stress on available resources may be it is space (land), water, minerals, energy or forest.

Usually the major causes of environmental degradation and overuse of resources are considered to be the population rise in the third world countries and their poverty but did ever we thought of the locus standing of family, kinship and community in this ordeal. Family, kinship, community sentiments and religion strengthen social bonds and social relations. When such ties are strong, it promotes virtues like cooperation, accommodation, social and emotional security etc. This ultimately leads to a social order in which the priority is with social life and economic pursuits as a mean proceedes. This elevates the moral fabrics of social life to a higher order in which not "self before others" but "Others before self" indicates improvement in the quality of life—the basic aim of development. Under such situation, the concept of an ecologically sustainable society can survive.

Conclusion

The development process has brought economic growth at the centre-stage by sidelining social and cultural issues. Its ramification is increased and marked inter-regional and inter-group disparities leading to a host of problems at the global, regional and local level. In the process, those socio-cultural variables which integrate the society into a cohesive unit have been denigrated thereby disturbing the symbiotic relationship between man and nature. Even the conception of sustainable development has not been able to encourage values of higher order which integrate the society and curtail the impairment of man-nature relationship. The symbiotic relationship can be established and strengthened through the use of common man's traditional knowledge and skills whose roots lie in strong familial and community ties. It will also encourage people's participation in the development process whereby higher order values (others before self, equality) will gain precedence over those which generate inequalities, disorganization and conflict. So long, those systems of social relations and cultural values, viz. family, traditions, norms etc. which integrate the society into a harmonious communities and neighbourhood continue to be ignored or relegated to second place, the idea of achieving a sustainable society seems, at least to me, a mirage.

REFERENCES

Aggarwal, Anil: "Toward Global Environmental Management', *Social Action*, Vol. 42, 1992.

Coser, L. A. *Masters of Sociological Thought*, Harcourt Brace Jovanovich, New York, 1971.

Dhaliwal, S. "Punjab Lets its Canals run dry", *The Tribune*, 24.01.2008.

Dodge, Jim "Living by Life: Some Bioregional Theory and Practice" in Dryzek and Schlosberg (ed.) *Debating the Earth*, Oxford University Press, New York, 1998.

Dryzek and Schlosberg (ed.) *Debating the Earth*, Oxford University Press, New York, 1998.

Harvey and Hallet: *Environment and Society*, The MIT Press, Cambridge, 1978.

Maio, Susannah "OPEC participation in 15th Session of UN Commission on Sustainable Development", *OPEC Bulletin*, Vol. xxxviii, No. 5, 2007.

Mitchell, W.C. and Simmons "Political Pursuits of Private Gain: Environmental Goods" in *Debating the Earth* by Dryzek and Schlosberg, Oxford University Press, New York, 1998.

Narain, Sunita "The Lifestyle Price", *Down To Earth*, Vol. 16, No. 17, 2008.

Raghunandan "Ecology and Consciousness", *Economic and Political Weekly*, Vol. xxii, 1987.

Sengupta, Samita : "Environment and Development" in Onkar Singh *et al.* (ed.) *Frontiers in Environmental Geography*, Concept Publishing Company, New Delhi, 1993.

Sharma, S.L.: "Sustainable Development" in K. Gopal Iyer (ed.) *Sustainable Development: Ecological and Socio-cultural Dimension*, Vikas Publishing House, New Delhi, 1996.

Sharma, Vibhu, "Climate Change put on top of World Agenda", *The Tribune*, December 27, 2007.

Sharma, Vibhu, "Greenpeace Raises Alarm on Automobile Emission", *The Tribune*, January 9, 2008.

Singh G., Anju and Poonam Rana, "Climatic Variability Impact on Human Health" in Jagbir Singh (ed.) *Environment and Development*, I.K. International Pvt. Ltd., New Delhi, 2005.

Virk, B.S. "Water and Power Crisis in Punjab", *The Tribune*, January 5, 2008.

WCED: *Our Common Future*, Oxford University Press, Oxford, 1987.

Yedla, S. "Environmental Valuation—An Effective Tool for Environmentally Sustainable Development" in *Proceeding of the Fourth International Conferences on Eco-Balance*, Japan, October 31 to November 2, 2000.

Yong, G.Y.V. "Lifeliness: An Indicator of Sustainability", in *Proceeding of the Fourth International Conferences on Eco-Balance*, Japan, October 31 to November 2, 2000.

————, "Rx: 60_minute Havan", *Down to Earth*, Vol. 16, No. 17, 2008.

5

Problems of Sustainability in Third World Countries

M.S. GILL

The concept 'sustainability' has a path to have economic growth based on policies that sustain and expand the resource base. It is a goal not only for the developing but also for the developed nations. The ideology of the concept has been based on the report published in 1980s—Our Common Future—predicting an ever-increasing environmental decay, poverty and hardship. The situation has remained the same to a large extent. The report of Third Global Environment Outlook indicates that by 2032, three per cent of the earth surface would become concrete, more than half the population would live with drought, 70 per cent of the remaining land and animals would be under threat and 16 billion tonnes of carbon dioxide would be released from fossil fuel annually (Global Outlook Report, 2002). This pained scenario makes one to think that despite talk of sustainable development, sustainability of the resources is far off. Further, sustainable society can continue for generations but the current development is not even closer to anything like sustainability. The degradation of environment and depletion of resource base of the world still continue to play truant to life on this planet and rapidly decreasing the prospects of sustainability. The question that needs attention, does Third World countries have the ability to make sustainable even at this stage of development? What are the basic problems faced by such countries that obstruct the goal of sustainability?

The economic and environmental indicators for sustainable development have been provided by the UN Department of Policy

Coordination and Sustainable Development (UNDPCSD, 1997). Some of the key factors of sustainability (Rao, 2000) are:

- An increase in the global warming potential,
- A net increase in the emission of greenhouse gases,
- The extent of deforestation especially of tropical and rainforest,
- Change in stratospheric ozone layer,
- Change in the public health features like incidence of skin cancer, epidemics and age structure-adjusted demand for health services for the upkeep of human well-being,
- Extent of loss of biological species,
- Changes in the incidence of weather extremities.

All these indicators are essential foundation for sustainability. But in reality, sustainability demands an adoption of life style within planetary ecological means which man has not been able to frame for him. An assessment of the type and magnitude of attributes of human activity explain the stresses on biophysical and ecological resources. Japan has chosen 20 headline indicators for sustainability based on an analysis of over 200 data sets in several sustainability—related categories and results show a decline of 19 per cent of sustainability since 1990 (JFS, 2005). Same may be the case with other countries and it is not difficult to imagine the situation of developing countries.

To keep pace with the rising population and the increasing demand for foodgrains, the farmers are made to use agro-chemicals to raise productivity and reduce losses due to pests, insects, and rodents, go for deforestation, multiple cropping, reclamation of land and the like. The increased demand for agro-chemicals is polluting our biophysical resources, depleting forest cover, and water resources. This is in addition to the chemicals used in industries. In India, additional 80 million tonnes of restricted pesticide products are legally released each year. These toxic chemicals are contaminating water, air and soil and getting diffused across the globe and produce synergic effect, thereby, greatly harming the life on this planet. Similarly, many a projects meant to ameliorate the conditions of the peasantry prove fatal to their aspirations and rather than improving their quality of life end up with converting their fertile fields to deserts

(Nanadkarni, 1987), transforming them into environmental refugees, or changing the very properties of soil like that in Punjab especially the Malwa area. In addition, the use of agro-chemicals leads to more incidences of asthma, bronchitis, skin diseases, and hypertension among those who handle these during transportation, storage and use. This not only increases their health expenditures but also diversion of funds from the productive use (Gill, 2003). On the other hand, the productivity of the soil gets affected due to use and retention of heavy binding compounds of the fertilizers and insecticides etc. Finally, the destined goal of achieving higher productivity on the farm leads to lower soil fertility and productivity. Is this process going to save the soil and its fertility for the benefits of the future generation? The World Watch, State of World 2000, report finds that increasing population pressure on ground water, croplands, oceans and forests is threatening the global environment. This coupled with the use of unclean technologies is furthering the effect by raising temperature. A study on the assumption that the level of carbondioxide in atmosphere will double from pre-industrial level, has reported that there will be a rise of earth temperature from 2 to 11°C (Nature, 2005 cited from OPEC, 2005). Technologies such as carbon sequestration and hydrogen fuelled vehicles among several other initiatives, of course are the answer to the challenges of pollution by greenhouse gases but to what extent are available or be implemented by the developed or developing nations, is yet to be seen. So, the *unclean technologies* and *population pressure* will not allow the developing countries in particular to achieve sustainability more so because their overriding priorities are economic and social development and poverty eradication. Further, introduction of new fuels including bio-fuel could be a good development but their production could compete for land, water and food which are also necessary for meeting Millennium Development Goal that is eradication of extreme poverty and hunger.

In reality, the pace of population growth can be slowed but not kept at par with mortality rates. Industrialization and urbanization will continue. Science and technological growth will keep its pace, lethal weapons will be produced for sale in large quantity, and man will go for the exploitation of available resources at the same rate and the like. This will take the planet away from the very conception of sustainable society. The solution in terms of new approaches and concepts may not solve the problem. It is in this context, this paper is an attempt to highlight some problematic areas.

The basic problem is the *fragmented approach at the international and national level.* The institutions facing challenges of development and resource use tend to act in an independent and fragmented way. These institutions are working at relatively narrow mandate with closed decision process. Those responsible for managing the natural resources and protecting environment are institutionally separated from those managing the economy. Yadagere village in Karnataka reported to the police about dumping of the industrial and bio-medical waste in the village but police did not bothered, it was only after six months when water in a borewell started turning yellow, the villagers got their water tested in laboratory and found strong contamination of aluminium and lead (Kaur, 2008). The waste was being dumped by the industrial units located in Tamil Nadu and crossed various borders without being noticed show the functioning of different departments. The Farm Security and Rural Investment Act, 2002 approved by US government has been an attempt to forsake environmentally sound policies for gain in domestic policies. Ironically, the European Union and official of WTO, IMF and World Bank have criticized the act. The policy promises subsidies that will benefit the agro-giants and multi-nationals at the cost of farming community and make environmentally harmful activities (excessive use of pesticides).The main purpose, as per the President George Bush was to sell beef, corn and beans produced in United States around the world at lower rates (Sethi, 2002). This is going to directly hit the agricultural and farm products produced in developing nations, as subsidies will lower the actual market price of the commodities. This shows that agricultural institutions works independent of the environment one and secondly, the rich nations are hitting the developing countries as the economy of latter overwhelmingly depend upon agriculture. The difference between the worlds's most affluent and its poorest can be seen in the contrast in nutrition, education, disease pattern, family size and life expectancy. This will cut into economic activity as funds, rather than investing in production process will be diverted to meet the basic necessities of the poor. As millions of poor are living in developing countries, their economy will be adversely effected not that of the developed ones. Thirdly, does the industries producing such toxic substances ever take the interest of consumers of primary products and that of the environment?

Also, citizens of wealthy countries share responsibility for ground water pollution and depletion of resources beyond repair. Eighty million Germans pollute the earth atmosphere with as many gases as one billion Indians do without realizing that water crisis is going to dominate the coming century and will lead to numerous conflicts (Rainer Horig, 2005). Further, despite adverse report by National Institute of Occupational Health, India, on the use of ENDOSULFAN on human health, industry lobby is coming out with reasons and reports in its favour. The industry has exploited the loopholes of the Pesticide Regulations (Yadav *et al.*, 2002). Not only this, the scientists of Agricultural University in line with the pesticide industry body, Agro-chemical Promotion Group, denounced NGOs for spreading rumours about pesticides despite high level of *aldrin, malathion, dimethoate and endosulphan* in water (Misra, 2007) and thereby having serious bearing on the health of the people. This implies that pesticide lobby without realizing the ecological or environmental consequences are supported by the very institutions meant for safeguarding the interest of the community and providing information and knowledge about the effects of use of pesticides, insecticides etc.

Another example will clear the mist. The Ministry of Environment in South Africa under the campaign of industry barons modified its ban on polythene bags by reducing their thickness from 80 micron to 30 micron with the plea that it would leave many unemployed. This shows how the interest of a few hinders the process of achieving the desired goal of sustainability and lead to environment hazards since most countries are trying to put a blanket ban on the use of polythene or prescribing stringent rules with regard to its quality and thickness. In the same vein, the thermal power plants emanate 500 m tonnes of ash everyday and it is usually dumped in ponds, marshy lands and agricultural fields affecting thereby peoples' surviving on these resources (*The Tribune*, 2000). To meet the energy demand of the people, industry and agriculture, the sources of subsistence of farming and tribal communities are encroached upon regularly. This displaces these communities and turns them into environmental refugees. Thus, it is not the institutions that are working in a fragmented way but at the same time these institutions are immune to the problems of the community.

Related to the above, it is observed that most of the rich nations/groups/individuals controlling and using the resources the most dictate terms and conditions to those who have no control over the resources. It is a known fact that developed countries comprising roughly 25 per cent of the world population consume 75 per cent of the world resources. Their control over resources generates *inequalities* which rather solving the issue of sustainability translates it into complex phenomenon. Even the global issue of "emission control" is not strongly supported by the developed nations responsible for global warming. Though Kyoto protocol has come into force on February 16, 2005, seven years after it was agreed, the world's biggest polluter, the United States, has not signed the treaty saying it would be detrimental to the country's economy (Webb-Muegge, 2005). The US attempt to thwart Kyoto Protocol on GHG emission makes one to belief that they want such imposition for the developing countries so as to restrict their economic growth in relation to that of the developed ones. G8 countries emit more than 40 per cent of the world's greenhouse gases. Moreover, the per capita emissions in developing countries are only fraction of those released into atmosphere by the developed nations (Griffin, 2007). So, why should the developing countries be asked to pay a price that would severely undermine their development prospects which developed countries have attained? In other words, rich are determined to ruin the developing countries either by imposing strict rules towards the use of resources through their control over monetary and technologically resources or defining priorities of growth or subsidizing their own products to control international market to their favour against the poor developing countries. Recently, a UNEP sponsored study found thick haze layers roughly seven times the size of India enveloping the north Indian Ocean (Singh, 2002). Such high concentration of pollutants was not earlier seen in the region. Scientists are worried at the fixation of aerosol; it could well be deadly for the regional climate—its rainfall and sunlight. Interestingly, greenhouse gases and regional haze are due to emission from fossil fuel burning.

Industrialized world is continuing to emit, far beyond its share, is exonerating itself and put the responsibility of such phenomenon on developing countries. It is said that cause of regional haze is the users of fossil fuel in developing countries without realizing their own responsibility for creating conditions for such environmental problem as the first explanation that differentiate developed from underdeveloped has been adopted from the west over-emphasizing the process of industrial growth. Moreover, climate changes in the first seven years have been occurring in increasing number of extreme weather events such as drought, flood and hurricanes. Such changes are alarming signals. So, we have to build a system in which more nations will work together under the common framework to stop global warming (Koizumi, 2005). This is an indicator that certain global issues are listed as regional or local issues and the international society remains a mute spectator to these without realizing the effect which will not be confined to a region; rather have its ramifications for the whole globe. The power of the rich it seems to be out of bound of the poor and that is why those countries are subjecting the poor to exploitative conditions of growth and prescribing stringent rules and regulation for them. China's economic prosperity has not been a pleasing development for US and its allies. In fact, in 2003-04 the US senate acknowledged that Chinese economic growth posed a major threat to US domination (Tane, 2008). It is through media US and its allies are making propaganda against Chinese products saying that Chinese made children clothing have high level of formaldehyde and this propaganda was found to be false after laboratory reliable re-testing. This shows that rich nations can go to any extent to further their interest as against that of the developing ones. Under such partisan conditions, the question of looking into indicators of sustainability as stated earlier and finding a long-term solution to environmental problems seems to be a far off cry.

Also, *materialistic culture and consumerist lifestyle* which has emerged over the years have precipitated the process of degradation of the resource base including land and forest. Mohanty (2005) finds that the consumerism of post-industrial societies has put natural resources under acute pressure. Farmers, industrialists, traders and rather all sections of the society are pursuing one and only one goal of having highest per unit production, highest per unit returns and income. With the result, factories and farms to have more production are overusing at least 40 per cent more of water reducing water level as in the case of downstream of yellow river in China (see Down to Earth, 2000), the wheat and rice rotation extensively followed in Punjab (India) has brought a crisis of water table which is receding at an alarming rate creating desert like conditions (Mahapatra, 1999). Despite provision of 2500 crores on the Ninth Five Year Plan, the available programme is unable to replenish our ground water resource. The water table in certain pockets of Amritsar, Jalandhar, Ferozepur, Ludhiana, Patiala, Sangrur and Faridkot has gone down by 4 metres in the last 20 years. Government though has framed laws to manage and conserve water but has no control over its production, usage or its sustainability. Even private dealers have come up who indulge in sale and purchase of water as a commodity in Gujarat and Rajasthan. As such the resources are ought to be mismanaged through overuse which may in the long-run lead to drought like conditions. Further, construction of large dams and opening of mines have destroyed local water resources and put villagers in trouble (Das, 2005). Here, it is important to quote Engels (1972) who said that the interest of the ruling class often becomes the driving force of production: since production no longer restricted to providing the barest means of subsistence for the oppressed classes and subsistence of dominated sufficient to ensure surpluses, they pay little attention to natural or social effect of these surpluses. In Rayagada district of Orissa—the home of Paroja-Kondha tribe—the private mining companies have got mining rights to extract bauxite which would eventually dry up around 700 streams and chew up the forest, the life of the district's 85 per cent tribal population (Kothari, 1996). This explains the condition of the developing nations where the ruling elite is robbing the resources of the poor who depend for their subsistence on nature. How can one think of even sustainable development what to talk of sustainability of the resources and society? Equally, the State Government instead of helping the poor who tried to improve the resource base, questions their efforts in this direction. The people were not allowed to build check dams in some places of Orissa, and at another level, the Government of Rajasthan tried to assert that the State has right over every raindrop and went out to demolish the water harvesting structures set up by the people (see Down to Earth,

2001). What a paradoxical situation, in one state private water owners are emerging and on the other hand the people trying to save whatever the resource they are subjected to state whims through exerted control over every raindrop of water. It is not the consumerist life style but the lack of political will to sustain the resource base through efficient management is another factor that hampers the process of growth with sustainability. It is required especially for the developing countries where ruling class finds one or the other excuse like loss of employment, development demand, industrial growth to catch up the west, colonizers demand to provide better housing (not to the poor but to the rich) an agenda of Government, and so on. Bromley and Cernea (1989) note the appearance of environmental management created through the establishment of Government agencies, and the aura of coherent policy by issuance of decrees prohibiting entry to and harvesting from State property has led to continued degradation of resources under tolerant eyes of Government agencies. It is not the dependent local communities but the outsider with the active convince of Government agencies degrade or deplete the resources. Also the failure of the Government to deal with falling water tables and the depletion of water aquifers could translate into food shortage, hunger among the poor and risk the confidence of the people in the Government potential to cope with such problems. It may lead to social breakdown and this type of situation has arisen in countries like Somalia, Afghanistan and the Democratic Republic of the Congo (Brown, 2005). With global warming and climatic changes coupled with depleted aquifers and exhausted soils, the number of environmental refugees are likely to swell into millions and Government if fails to take actions with political will, may not survive economically and socially.

Meadows *et al.* (1972) assume that each additional person would need 0.08 hectares of land for purposes like housing, roads, waste disposal, power lines and other uses and 0.04 hectares for producing the food he needs. With the limited land available at disposal, the rising population will ultimately encroach upon the land that poor have. In the process, the poor will continue to be displaced from their sources of subsistence, turned into environmental refugees; arable land going for industrial, mining and housing use; in addition to the non-judicious use of the available resources including land, water and forest and the like. How can then sustainability of resources be achieved? Developing nations cannot afford the deterioration of the productivity of the soil and change of land use. But it requires decisions with regard to matters related to soil, land use, type of technological adoption and its ramification over the years. The issue of Special Economic Zones (SEZ) to be developed in different States of India is racking up with farmer lobbying against it as it will devoid them of their source of livelihood despite Government promises of handsome compensation. The events in Nandigram (Kolkata) turned worse as farmers to a large extent are against their cultivable land to be handed over to industrial lobby. The root of such issue lies in forces of power, privileges and property relations. The struggle for these not only differentiate the society but leads to conflict and aggression/suppression. These eventually disturb the functioning of various institutions meant for governance.

The process of development with economic growth and the pattern of access to and the use of natural resources are leading towards an ecologically and socially impaired society. In the whole issue, erosion of social and cultural values never appears. The erosion of these values has undermined the system of family, community and neighbourhood. There is inverse relationship between the level of development and of family stability, neighbourhood sentiments, and community feelings (Sharma, 1996).These institutions have worked in the past to attain the highest form of value system in terms of equality, fraternity, man-nature relationship and in spite of contradictions and conflicts with the active support of religious doctrine and cultural ethics. There has been a complete disruption of social support of networks. Earlier, the families use to enter into an exchange system where reciprocal labour sharing arrangements existed. These arrangements were based on kinship relations and extended beyond the village on kinship. Such arrangements provided social and economic support as well (Eglar, 1960) besides extending strong kinship bonds for the welfare of community. Disruption of such support system has isolated the individual from the community, disintegrated the joint family system and has given way to individualistic tendencies. At the cultural level too, the economy has come to occupy the central stage and hierarchization of culture has taken place (Nandy, 1989). Under these circumstances, a perpetual sense of subjugation comes to form the ideology of the people and this may lead/has led to an atmosphere where all aspects are taken to be under threat and need to be protected jealously and in fact, this may be the root cause of emergence of militancy and build up of war like situations. It is out of this contention; the concept of identity recognition has been added to the definition of development.

Some traditional religious and folk beliefs also contributed to the preservation of nature especially (Gadgil and Vartak, 1975). Moreover, the shift from community control and management of common property to state or individual ownership and control has increased environmental degradation (Dasgupta and Maler, 1990). In this way, individual pursuit to growth has achieved but the common property sources have been either destroyed or depleted. Punjab is an example where common agricultural property (*Shamlat* Land) has been ushered by the private owners to the utter neglect of community right over it and this has intensified the struggle between the individuals for control over such resource. In addition, there has been a systematic devaluation and marginalization of indigenous knowledge about species-varieties, natural processes (how forests, soils and water are formed and sustained interrelatedly), and sustainable forms of interaction between people and nature (Rangarajan, 2007). Similar problems are emerging in China (Glaeser, 1987). Along with, the hierarchization of knowledge into traditional and scientific has eroded knowledge developed through the interaction with nature as demeaning.

To have a sustainable society and to conserve environment and ecological balance, it is necessary to raise humanity from narrow interest to the higher ones. The lust for uncontrolled use of natural resources, tendency to have more and more are manifest of generating conditions of inequality, which eventually put the economic variables above that of social ones, thereby leading to further degradation of the environment and loss of biodiversity besides translating this struggle into war over the control and use of resources. The gulf war to a large extent was the manifestation of the above issue. Rural households in India like that in most of the Asian and African countries gather their everyday personal use and sale, such as food, fuel, fodder, fibre, bamboo, herbs etc. from the village commons and forests and as such their conservation and management should be entrusted to the villages rather than to private or Government agencies. Their access to village commons reduce income inequalities to some extent as the dependence of the poor is especially high for fodder and fuel: village commons supply more than 91 per cent of firewood and more than 69 per cent of their grazing needs, compared with the relative self-sufficiency of the larger landed households (Aggarwal, 2007). Even the class differences will reduce if ruralites are allowed to maintain forests responsible for the fertility of soil and availability of water (both for irrigation and consumption), the lifeline of rural society.

Quoting Gandhi here will not be out of context. Governments and States in modern times try to regulate individual behaviour through fiscal and monetary regulations, try to control activities that in their opinion are not in the interest of the society. Gandhi wanted that such control should start from within the individual as external factors undermine individuality and internal constraints uplift the individual (cited from Kumar, 2002). If this ideology becomes the part of the individual, the issue of sustainability will come to cease to a large extent. Further people's inclusion in decision-making process will enhance transparency in governance, accountability for governmental actions for the preservation of environment and their involvement in process of development and growth. The villages which have harvested the raindrops through community action have fought drought against those who could not do so. These villages have been able to generate income with bumper crop even under drought like conditions (Das, 2002). Moreover, instead of building large dams that inundate large area of cultivable and forest land, displace population and irreversibly altering the ecosystems, stress should be laid on constructing earthen dams with local material and labour to collect and store rain water. Likewise the people of fragile land have got its own traditional social capital, human capital and management of these assets and capital resources may help people to come out of poverty. This shows that if the individual realizes the importance of community action to sustain community life, it will restrain individual lust for control over the resources. Similarly, if joint family system gets priority over the nuclear one, the demand for a number of consumer goods will automatically fall down putting less strain over the resources.

Conclusion

The concept of 'sustainable development' in the present modern society has not been able to find a place in the agenda of development and growth, as the rich nations are more concerned about the welfare of their own people rather than that of the humanity at large. Further, these nations are creating unequal conditions for developing nations with their control over most of the resources and the multilateral agencies. Not only this, these nations do not share technological advancements with the developing nations. These nations either have to rely on the obsolete technological information provided by the rich nations or have to part with a big chunk of their budgetary provision which otherwise could be used for providing a quality of life. Majority of the developed nations consider third world countries as their dumping grounds. Secondly, the fragmented approach on the institutions managing environment and economy in no way support the sustainability of resource base and sustainable development. Thirdly, the quest for control over resources of the poor to meet materialistic demands of the rich does not help in the process of sustainability. No doubt, lists of indicators that support sustainable development to achieve sustainability have been provided but the lack of political will and determination and fragmented approach hinder the process of attaining the desired goal.

Only solution in vogue is to have one's mind and also in the heads of the State the attitude NOT ME BUT YOU and OTHERS BEFORE SELF. This will sustain cultural and social fabrics besides achieving sustainable society. In these circumstances, the ideology of Gandhi serves an important base for achieving the objective of sustainability, which in my opinion is a utopia. Realizing man as a social being has sentient relationship with material things, the present modern society need to indoctrinate the following ideas in its endeavour to achieve sustainability:

- Self-discipline in the use of resources,
- Synthesis between what is being preached and practiced by the States of the world,
- Dignity of humanity over self-assertion through internalization of familial and community values, and

- Making a space for the growth of underprivileged and underdeveloped sections of the society and States.

REFERENCES

Aggarwal, Bina "The Gender and Environment Debate: Lesson from India" in Rangarajan (ed.) *Environmental Issues in India*. Pearson Education, New Delhi, 2007.

Aims of Establishing Sustainable Indicators, *Japan for Sustainability (JFS)*, www.japanfd.org/en/view/ 9/7/2005.

Bromley, D.W. and M. Cernea "The Management of Common Property Natural Resources", *World Bank Discussion Paper No. 57*, World Bank, Washington DC, 1989.

Brown, R.L. *Plan B: Rescuing Planet Under Stress and a Civilization in Trouble*. Orient Longmen Pvt. Ltd., New Delhi. 2005.

Das, Binayak *et al.* "A Midsummer Dream", *Down to Earth*, Vol. 11, No. 3. 2002.

Dasgupta and Maler, "The Environment and Emerging Development Issues" cited from Rangarajan (ed) *Environmental Issues in India*, Pearson Education, New Delhi, 2007.

Eglar, Zekiye *A Punjabi Village in Pakistan*. Columbia University Press, New York, 1960.

Engles, F. *Dialectics of Nature*, Progressive Publishers, Moscow, 1972.

Gadgil and V.D. Vartak, "Sacred Groves of India: A Plea for continued Conservation" 1975 cited from Rangarajan (ed) *Environmental Issues in India*. Pearson Education, New Delhi, 2007.

Gandhi, M.K. in S.B. Kumar, *Environmental Problems and Gandhian Solution*, Deep and Deep Publications, New Delhi, 2002.

Gill, M.S. (ed) *Punjab Society*, Concept Publishing Company, New Delhi, 2003.

Glaeser, B. (ed) *Learning From China? Development and Environment in Developing Countries*, Allen and Unwin, London, 1987.

Griffin, James, "The Planet and Its People", *OPEC Bulletin*, Vol. XXXVIII, No. 8, 2007.

Kaur, Ravleen "Toxic in your Backyard". *Down to Earth*. Vol. 16, No.16, 2008.

Koizumi in "Kyoto Protocol comes into Force" by Webb-Muegge, *OPEC*. Vol. xxxvi, No. 3, 2005.

Kothari, S. "Whose Nation, the Displaced Victims of Development", *Economic and Political Weekly*, January 15, 1996.

Mahapatra, R. "Punjab Facing Groundwater Crisis", *The Tribune*, October 28, 1999.

Meadows *et al. Limits to Growth*, Universe Book, New York, 1972.

Misra, S.S., "Toxic Nexus", *Down to Earth*, Vol. 16, No. 15, 2007.

Mohanty, M. "Urban Poverty, Energy Use and Environment/Health Implications in Small Islands Developing States" in Singh (ed.) *Environment and Development*, I.K. International Pvt. Ltd., New Delhi, 2005.

Nandkarni, M.V. "Agricultural Development and Ecology", *Indian Journal of Agricultural Economics*, Vol. 42, No. 3, 1987.

Nandy, A. in John S. Augustine (ed) *Strategies for Third World Countries*, Sage Publications, New Delhi, 1989.

Rainer H. "Harvesting Water", *German News*, Vol. xlvi, 2005.

Rangarajan, M. (ed) *Environmental Issues in India*, Pearson Education, New Delhi, 2007.

Rao, P.K. *Sustainable Development*, Blackwell Publishers, Oxford, 2000.

Sethi, N. "Costly Largesse", *Down to Earth*, Vol. 11, No. 3, 2002.

Sharma, S.L. "Sustainable Development: Socio-cultural Imperatives" in Iyer (ed) *Sustainable Development*, Vikas Publishing House, New Delhi, 1996.

Singh, Neelam "Aerial Raid", *Down to Earth*, Vol. 9, No. 13, 2002.

Tane, Haikai "West Worst", *Down to Earth*, Vol.16, No. 16, 2008.

Webb-Muegge, "Kyoto Protocol comes into Force", *OPEC*, Vol. xxxvi, No. 3, 2005.

Yadav *et al.* "Endosulfan Conspiracy" *Down to Earth*, Vol. 11, No. 4, 2002.

———— "Thermal Plant Pollution hits 30 Villages", *The Tribune*, September 29, 2000.

———— "The South", *Down to Earth*, Vol. 9, No. 13, 2000.

———— "Who owns the Rivers", *Down to Earth*, Vol. 10, No. 4, 2001.

———— "Global Environment Outlook 3", United Nations quoted from *Down to Earth*, Vol. 11, No. 3, 2002.

6

Development of Science and Technology
Impacts on Human Security and Survival (A Sociological Analysis)

JASLEEN KEWLANI

Introduction

Discourses and discussions of science and technology are always undertaken in the framework of 'objectivity' as it is largely believed that all scientific events take place out of the rubrics of bias. But an abstract and even occasionally concrete contradiction can be discovered in the statement. On one hand, where science is assumed to be a magnificent source of objectivity, there only the related developments have forwarded enumerable deprivations and differences to specific segments of demography. These adversities in many instances result into human insecurity. Conceptualizing 'Human Security' or the thereby posed threats in a tedious task as it involves multi-dimensional analysis of multiple issues of life. The concept of human security is especially relevant for developing regions like South Asia as they are still grappling with post-industrial political-intellectual issues like nation-state formation that the developed nations have succeeded in vanquishing over their long periods of political evolution. When the term industrialization is being used, it of course, implies mechanization and automization; which again reminds the humanity of its relative deprivation. As a matter of serious concern, the danger to human and State security apropos becoming less amenable to state management in South Asia due to the growth of transitional and sub-natural challenges to its authority. A further erosion of human security arises from global factors like environmental degradation; hence the possibility of conflict is high because the resources that are necessary to support even the current populations are fast becoming scarce (Chari and Gupta, 2003). At this point of discussion, it is better to describe the concept of human security than to define it. The UNDP has called for a transition from narrow concept of national security to the all encompassing concepts of human security (UNDP, 1994). It also distinguishes between the perceptions of rich and poor countries. The rich seek security from the threat of crimes and drug, wars in their streets, the spread of deadly diseases like HIV/AIDS, soil degradation, rising levels of pollution, the fear of losing jobs and disintegration and deterioration of social fabrics. On the other hand, the poor nations demand liberation from the continuing threat of hunger, disease and poverty. The criteria of these generalization has not been yet classified but threats to human security have been placed within seven categories, viz. economic security, food security, health security, environmental security, personal security, community security and political security (UNDP, 1994). The present paper intends to elaborate the impact of scientific and technological developments on the human security encapsulating some important dimensions of the concept. Humanity is secured if the adopted developmental paradigm encircles the aim of sustainability.

There are some aspects of existentialism, which are invested due to development of science and technology and on the other hand, rest of the aspects are rendered as adamant deprivation. These deprivations forward pernicious restraints to sustainability, due to which the letters of development attain an evanescent form. Sustainable Development can be achieved if all the chores of survival, viz. economical, political, infrastructural, environmental, educational, etc. are equally emphasized. If any of these is divested of an adequate maintenance, the aims of sustainability become success proof. Another credential of sustainability implies that the needs of the present generation should be fulfilled in such a way by using the available natural resources that the future generation is not limitized to use the same natural resources when it needs them. Paradoxically, development of science and technology has produced constraints for sustainable development. Environmental degradation due to rampant industrialization; increasing female foeticide with the misuse of pre-natal diagnosis techniques; agricultural crisis due to inflationary pressures of technological devices and machines required for an efficacious agricultural practice; and ill health due to pollution increase; are some of the adversities entailed through incessant technological development. Honestly, the advantageous creations produced by science and technology for mankind can also not be ignored. Whenever the causes behind enhancement in standard of living are analyzed, science and technology come at the apex of the pyramid. Electronic gadgets, motorization and automobile industry, development of communication with the help of cell phones, diagnostic techniques, information technology and many other inventions and discoveries have contributed to the prosperity and contentment of mankind. These bi-dimensional effects of science face the controversy. One group of academicians, intellectuals and commoners profess that science and technology have laminated human life into a safer cover which facilitates a comfortable platform to survive. On the other hand, the other group counters by holding an 'ism' against the development of science and technology. It believes that humanity has been trapped in vicious inhibitions due to science and technology and ample adverse effects have been showered by scientific and technological innovations, which are complex to be digested. This paper aims at analyzing the multiple ramifications of development of science and technology; which have been encapsulated in the fabrics of contemporary conceptualization of development.

The Contemporary Conceptualization of Development

Mere economic interpretation to the term 'development' has now been abandoned. The concept of development has acquired multifaceted form for an absolute understanding. Till the time only economic development was targeted; development of science and technology was considered to be a boon. But now, when Human Development and Social Development have acquired an equal importance to that of economic development, some events related to science and technology are disdained. Individual as well as societal domains of human life need to be stressed upon simultaneously, if Social and Human Development are aimed to be achieved. The concern for human development is an attempt to rectify the failures of the prevailing dominant development approach, which had resulted in skewed distribution of resources and opportunities. It was realized by the 'Developmental Scientists' that the fundamental goal of development policy should be to create sustainable improvements in the quality of life of people (Pant, 2006). But as has been admitted in the Human Development Report, 2002 of UNDP, the human development approach to development has fallen victim to the success of Human Development Index (HDI). HDI attempts to capture the domain of development via the indicator of decent standard of living, it fails to address the problem of social, political and economic deprivation completely. Somewhere or the other, the rising standards of living of individuals involves technological and scientific enhancements as an indispensable criterion.

Political freedom, economic independence, social opportunities, administrative transparency and above all an adequate arrangement of social facilities like education, health and career, can together pave the way for holistic development rather than only steps in addition to technology. If the adversities forwarded by the changes occurring due to economic and scientific expansion are not tackled with, there may occur a risk of social unrest compounded with a divested survival and an unstable administration. 'Millennium Project' of 'The Awareness Council' of United Nations University, made a comparative analysis over the positive and negative ramification of development of science and technology. Information technology possesses a potential to operationalize the term 'Global', but it does intensify the impulses of fraud, cyber-terrorism, information warfare, cultural threats, etc. According to its report, low cost methods to determine the sex of child can drastically affect the demographics of humanity and no doubt animal cloning improves quality and cut cost of animal products but it raises ethical questions about potential of human cloning (Ginkle, 2003). Hence, development of science and technology is affecting the human life from both the angles, i.e. negative as well as positive. Though economic development is being achieved with the help of scientific innovations and discoveries, still, human and social development are retarded intentionally or unintentionally.

Categorical Influences of Technological Development

Scientific and technological activities in India are carried out under a wide set-up consisting of Central Government, State Governments, higher educational sector, public and private-sector industry and non-profit institution and associations. These institutional structures along with their research laboratories are the main contributors to Research and Development (Government of India, 2006). Some notable institutions among these are Council of Scientific and Industrial Research (CSIR), Indian Council of Agricultural Research (ICAR), Indian Council of Medical Research (ICMR) and many more. Further, there are over 1,200 in-house research and development units in industrial undertakings supporting research in their respective areas of functioning. In the following course of the discussion, various issues will be discussed which pose a compulsory link with technology.

Industrial Growth and Development

Industrialization is the most eye catching phenomenon of scientific and technological development, being the strongest indicator of economic development. A fact remains intact that growth of industrialization leads to the growth of environmental pollution by all means. Industries are established by replacing some or other ecosystem's units like forests, valleys etc. Industrialization leads to deforestation, which further leads to increase in environmental pollution. Firstly, because the green belt is attacked for economic purpose; and secondly, industries result in emission of poisonous gases into open air. Growth in the industrial sector in 2003-04 continued to be healthy. The index of industrial production (11P) grew by 6.9 per cent. Table 6.1 shows Annual Growth Rate of industrial production of major sectors of industry.

The major element of the buoyancy the industrial growth was the manufacturing sector with 80 per cent weight in IIP. Manufacturing witnessed a growth of 7.2 per cent, while mining and electricity grew by 5.1 per cent, and 5.0 per cent respectively (Economic Survey, 2003-04). With this rampant growth of industrialization, pollution occupies the place of the most obvious

Table 6.1: Annual growth rate of industrial production in major sectors of industry (based on index of industrial production) Base 1993-94=100

Period	*Mining and Quarrying*	*Manufacturing*	*Electricity*	*Over all per cent*
Weight	10.47	79.36	10.17	100.00
1994-95	9.8	9.1	8.5	9.1
1995-96	9.7	14.1	8.1	13.0
1996-97	-1.9	7.3	4.0	6.1
1997-98	6.9	6.7	6.6	6.7
1998-99	-0.8	4.4	6.5	4.1
1999-2000	1.0	7.1	7.3	6.7
2000-01	2.8	5.3	4.0	5.0
2001-02	1.2	2.9	3.1	2.7
2002-03	5.8	6.0	3.2	5.7
2003-04	5.1	7.2	5.0	6.9

Source: Central Statistical Organization, 2005.

repercussion. Pollution control has assumed a significant role, and commensurate with industrial expansion, as strong regulatory mechanism is required to ensure minimum environment standards to achieve the objective. A total number of 2,155 pre and post-1991 units of 17 categories of highly polluting industries have been identified for this purpose by Central Pollution Control Board.

Out of these units 1,877 units have been provided adequate facilities to comply with the standards, 225 units have closed down and the remaining 53 units do not have adequate facilities to comply with the standards. Legal actions have been initiated against all the defaulting units. At the same time industrial growth has been at below 10 per cent for several years now. One of the most important challenges in Indian economic policy consists of deriving strategies for obtaining industrial growth. Though this aim caters to the needs of scientific and technological advances but it may further reinforce environmental degradation.

Automobile Industry and Air Pollution

Industries and automobiles are by far the main contributors to air pollution across the world. Other than industrialization, development of automobile sector is another impact of scientific and technological development. The automobile is a symbol of modern science and technology and is indispensable for transporting men and goods. But some environmentalists hold automobile industry as a culpable factor for environmental pollution. They believe that automobile is the worst invention of human mind, perhaps next to the nuclear bomb (Rajagopalan, 2005). Bringing up a critique of private transport will be an injudicious step because in modern times of consumer activism, the cardinal need of each individual is to own his own private vehicle. Demand for private transport seems insatiable. From about 50 million in 1950, the number increased to 520 million in 2002. Not only India, but the automobile industry dominates economy of many nations. The US alone has 25 per cent of all the cars in the world. Developing nations like India and China actively promote the production and private ownership of cars. The evidence is presented in the easy and convenient loan provisions available in private as well as public banks of the nation. Where automobile development is an indicator of technological upliftment, there only it contributes to a range of environmental problems: increasing air and noise pollution, adding to solid waste, accelerating global warming, taking a heavy toll on human life through accidents, using up natural non-renewable resources like oil and metals, etc. The use of the automobiles further causes lead pollution which is a pernicious problem. Lead was in fact added to petrol to prevent 'Knocking' in the engine which is however extremely poisonous; especially when it gets accumulated in most of the biological system. Excessive accumulation of lead in the body of a being can result in paralysis, blindedness, and even death. It also devastatingly effects the mental development of an unborn child. These are the reasons for why most of the countries have banned addition of lead in petrol. This step is a cogent need in lieu to control pollution because with an incessant increase in private transport, air pollution is also increasing in India.

New Delhi has been rated as the forth most polluted city in the world by World Health Organization (WHO). However, Delhi is not at the top of the list of polluted cities. India has several other pollution hot spots also: Central Pollution Control Board (CPCB) released a report in 2003 and stated that Ahmedabad's air is the most noxious; followed by Kanpur, Sholapur and Lucknow with small particulate levels that is 3-4 times the standards of 60 microgram per cubic metre (mg/m^3). The report has ranked 29 cities according to the Respirable Particulate Matter (RPM) levels recorded during the year 2000. This report thus confirms the fact that Indian cities show high particulate pollution, with 14 cities hitting critical levels. Nitrogen dioxide levels in most major cities are generally closer to the acceptable annual standard of 60 mg/m^3. These intense statistics are the results of heavy vehicular traffic and density, especially in case of few locations situated in Kolkata and Delhi. Another alarming fact is that residential locations in India are fast outpacing industrial locations in air pollution and the vehicular fumes are responsible for this verbatim trend of air pollution (Bharucha, 2005).

It will not be unjust to say that the dominant reason for higher levels of air pollution in India is the development means of transportation. Technology has facilitated human being with uncountable types and models of vehicles suitable to all the economic classes of the world. Some manifest drivers of air pollution are: high vehicular density in urban areas; poor and unregulated fuel quality; predominance of two stroke vehicles; lack of efficient mass rapid transportation systems; uncontrolled growth of vehicular population; improper traffic planning inadequate maintenance of vehicles; inadequate pollution prevention and control systems and lack of enforcement of pollution standards. A large number of two-stroke vehicles (up to 70 per cent of total vehicle population) run on the roads of NCR Delhi. Rampant adulteration of petrol/diesel makes the situation worse. Numerous small diesel run generator sets (in excess of 150 thousand in Delhi) spoil the urban air quality because of frequent power failures. Further, many small scale industries use furnace oil with more than 3.5 per cent sulphur content. A number of measures have been adopted by the Government to control the vehicular pollution, which include introduction of stricter emission norms (Euro/Bharat norms). Fuels quality has been improved significantly and since 2000, metro cities are supplied with unleaded petrol instead of leaded petrol and 0.05 per cent sulphur content diesel (since 2003) in place of 0.5 per cent sulphur content diesel. All diesel-operated buses have been converted to operate on CNG in Delhi. The improvement is ocular in fewer cities only. Generally speaking the condition of urban environment is deteriorating in an unhesitant pattern. The evidence for the transport driven air pollution is that the fuel demand for sustaining urban transport has grown several folds in last decade and this rate is escalating further and putting additional pressure on the foreign exchequer. By 2050, world population will be approximately 8-10 billion with 80 per cent people living in urban areas. Their average income will be in the range of US$15,000 to US$25,000 per annum. The per capita energy demand in 2050 will be 2-3 times of the present level. The relationship between per capita income and energy requirement is shown in the figure given below. Analyzing the situation presented with the help of this figure it can be said that India faces a great challenge at present. It needs to follow a flat trajectory of growth in fuel demand. Alternatives have to be considered in order to undertake import substitution for diesel and petrol fuels.

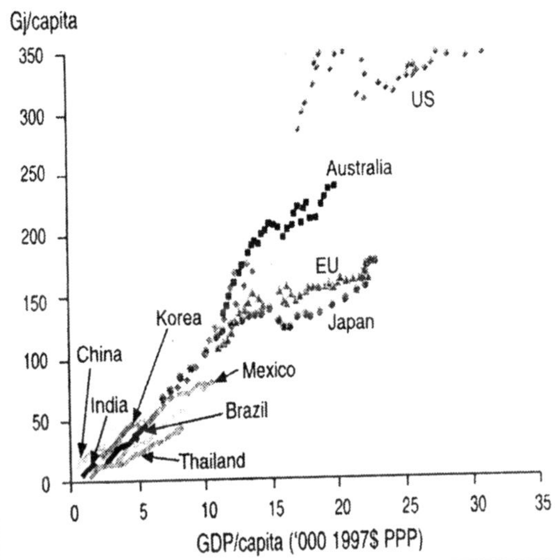

- +$25k/capita: little extra energy needed.
- +$15k/capita: services start to dominate growth.
- +$10k/capita: industrialization near complete.
- +$5k/capita: industrialization and mobility take off.

Fig. 6.1: Climbing the Energy Ladder: A Continuously Changing Relationship

Source: India Infrastructure Report, 2006.

No single fuel can sustain urban transport in foreseeable future. This excessively increasing demand of fuel directly implies an increase in air pollution, hence, it is justifiable to assert that development means of transportation and air pollution goes hand in hand. Most prominent reason of air pollution is motor vehicles. Suspended Particulate Matter (SPM) levels in six major metropolises is well above the 'National Ambient Air Quality Standards'.

Table 6.2 shows the Air pollution levels of prominent metropolises of the nation.

Table 6.2: Air pollution levels

City	*SO_2 (µgm/cu.m)*			*NO_2 (µgm/cu.m)*			*SPM (µgm/cu.m)*		
	1993	*1998*	*2003*	*1993*	*1998*	*2003*	*1993*	*1998*	*2003*
Delhi(Nizamuddin)	13.70	15.60	12.20	30.10	35.10	43.30	362	342	315
Mumbai (Bandra)	49.50	15.90	7.70	32.30	14.70	18.70	475	211	219
Kolkata (Lalbazar)	65.10	47.20	18.0	62.00	39.70	75.50	507	283	244
Chennai (General Hospital)	10.30	10.30	6.60	27.10	15.40	7.50	73	131	149
Bangalore (Anand Rao Circle)	—	41.60	10.80	—	28.40	44.90	—	239	198
Hyderabad (Abids)	7.30	7.60	9.70	11.00	22.10	19.50	156	152	139
National Ambient Air Quality Standard (Residential areas: annual average)		60			60			140	

Source: Compiled from data contained in the website of the Central Pollution Control Board (http://www.cpcb.nic.in)

The pollution levels depicted through this table are a threat to sustainability and to human survival also. Industrialization has also contributed to this scenario of environmental deterioration. Landfill sites also release landfill gas with 50 to 60 per cent methane by volume. Methane is 21 times more potent than carbon dioxide aggravating problems relating to the issue of global warming. It is estimated by the Energy and Resource Institute (TERI) that in 1971 India released about 7 million tonnes of methane into the atmosphere. This could increase to 39 million tonnes by 2047 if no efforts are made to reduce composting, recycling, etc. (India Infrastructure Report, 2006).

Water Pollution and the Affiliated Crises

At the start of 21st century, waterborne infections diseases are a thing of the past rich countries, accounting for a fraction of 1 per cent of overall mortality. At the turn of 19th century, diseases like diarrhea, dysentery and typhoid fever posed major threats. In the late 19th century they accounted for 1 in 10 deaths in US cities, with children the primary victims. For people in the rich country it is difficult to imagine what water insecurity means in developing countries. Falling reservoirs, declining rivers, hosepipes bans and political exhortation to use less water are becoming more common in parts of Europe. In United States management of water shortages has long been a public policy concern in States such as Arizona and California. But almost everyone in developed world has safe water available at the twist of taps. Almost nobody dies of want of clean water or sanitation—and young girls are not kept home from school to fetch water from the natural water bodies or resources. In contrast of this, position of the developing world is entirely different. The situation of the availability of water and its quality thereon is a morose in the developing world (Human Development Report, 2006). The condition of availability of water and its quality has been further deteriorated with development in technology in form of industrialization because industries pollute water bodies by exposing the solid and semi-solid wastes into them. Studies throughout the world have proved that where industries have to self provide this natural resource, costs of production go up sharply, competitiveness is reduced, and economic growth is damped. The 'self-provision' of water supplies is just one manifestation of a far broader breakdown of public infrastructure in India. A recent survey shows that 60 per cent of Indian manufacturing entities have captive power generating units—a figure which is just 16 per cent for China, 17 per cent for Brazil, and 42 per cent for Pakistan. This groundwater revolution brought immense benefit to India, playing a major role in the irrigation/rural development/poverty reduction, achievements. It is now ocular that the ground water revolution has run its course in the most productive agricultural and urban areas of the economy; as a consequence, there emerges a major sustainability challenge. This challenge is pertinent to the resource itself i.e. "water". Average figures of water availability show that the annual replenishable, groundwater resources of India amount to about 430 billion cubic metres (bcm), and that net withdrawals amount to about 160 bcm per year. There would, therefore appear to be little problem 'on averge'. But in fact, all water issues are local issues, and averages flatter to deceive. At local level many of the most highly productive localities are already under severe groundwater stress. For example, in Punjab groundwater in about 60 per cent of the blocks is either already being, or very near to being, overdrawn, while for Haryana and Tamil Nadu the figure is already around 40 per cent. The following figure shows the state of groundwater in Tamil Nadu.

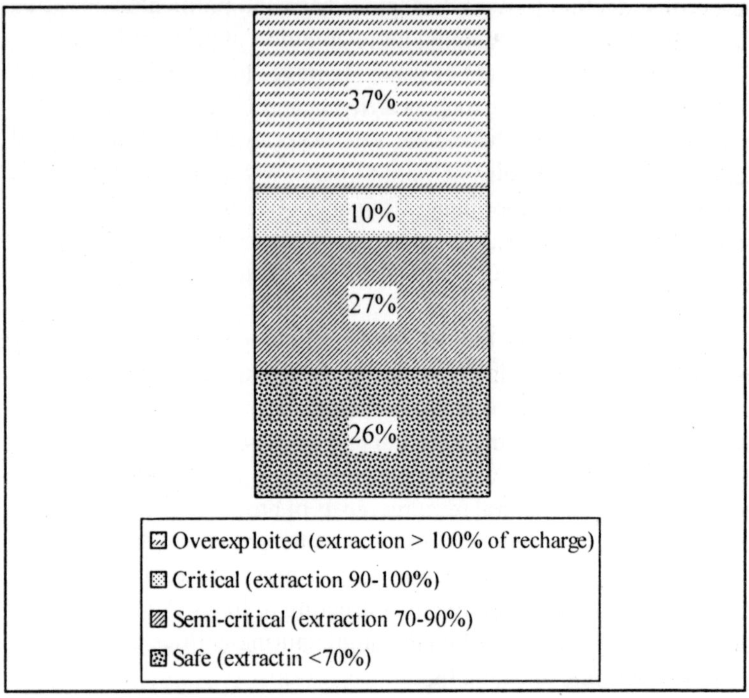

Source: See Briscoe, John, 2007.

Fig 6.2: The precarious state of groundwater in Tamil Nadu

The state of water availability is not only pathological in Tamil Nadu only; rather in mostly all the States of the nation. In Rajasthan, the proportion of overexploited blocks has risen from 17 per cent to 60 per cent over the last seven years. For the country as a whole, about 14 per cent of all blocks are either over-exploited or critical. Hence, water related challenges have been aggravated with technological development in the form of industrialization, construction of Dam, developed irrigated modes, etc. (Briscoe and Malik, 2007).

Industrialization has spared no chore of environment from being degraded. Water contamination is another ocular result of industrial development. The profligate consumption of oil and coal is only one of many pollution threats from industrial processes currently facing our fresh water resources. Today's pulp industry uses 60,000 to 190,000 gallons (1 gallon = 3.78 litres) of water per ton of paper or rayon. Bleeching uses 48,000 to 72,000 gallons per ton of cotton (Climate Change, 2001). The amount and proportion of water used in manufacturing is rising as industrialization speeds up around the world. A simultaneous phenomenon is that a huge quantity of industrial effluent is discharged into freshwater ecosystems. This quantity has been rapidly rising from 237 billion tons in the mid-1980s to 468 billion tonnes by 2000. Prior to human beings the contamination and paucity of water affect the other largely dependent species of the world; creating an irreparable loss to the universality of the biodiversity composite of mankind also. Widespread concern is now routinely expressed for the degradation suffered by inland water. Worldwide (including India also), of the creatives associated with inland waters, 24 per cent of mammals and 12 per cent of birds are severely threatened, as are a third of fish species (United Nations, 2003).

Development of science and technology in the form of industrialization plays havoc with the ordinary masses who claim a natural possession over the natural resources around them. In a landmark case of India involving Coca Cola, the company was recently ordered by the High Court in the Southern State of Kerala to stop drawing on a groundwater aquifer. Villagers claimed that its bottling plant was drying out the land, destroying their coconut palms and turning their rice paddies into a dust bowl. In December 2003, Justice K. Balakrishnan Nair found in their favour and delivered a statement, "Groundwater under the land of the company does not belong to it, every land owner can draw a reasonable amount of water necessary for domestic and agricultural requirements. But here, 510,000 litres is extracted per day, converted to products and transported, thus breaking the natural water cycle." This particular litigation highlights the agricultural crisis derived as a consequence to industrial development. But it is not the only way by which agriculture is haunted due to technological advancement. The agricultural community has also not been spared by the influences of scientific development. This part of the discussion will be given a place in the following course of the paper.

The newly invented techniques and methods for irrigation and cultivation are very expensive. When these inflationary pressures adulate with the Government's fiscal and economic policies for agrarian society, it showers alarming effects. Technological instruments and machines are not easily adopted and bought by the farmers for which this category adhere to loans being provided by the private agencies on exaggerated rate of interest further traumatizing the agrarian community. For repayment, they borrow more loans which unfortunately develop into a web of borrowance. This vicious circle of loans traps the farmers. Due to this phenomenon, 'suicide' has become a very prominent and often occurred in agrarian communities. In all, it will not be a myth if said that atomization of land holding and mechanization have been major causes of agrarian crises specially in some agriculturally dominated States like Punjab (Punjab Human Development Report, 2004). Not only physical well-being of humans is constrained, rather they are deprived of a healthy environment which is very crucial for a disease less survival. Suffocation, asthma, brain tumor, lung cancers, etc. are some of the chronic diseases which come out as a possible effect of industrial pollution. All this needs to be managed because introduction of technologies in energy production and transport, fuel switching, increasing transportation efficiency (with effective pollution control) and promotion of investment in public transport are some resorts which may pacify the ongoing and uncontrolled pollution expansion (UNEP, 2000).

Solid Waste Disposal and Environmental Management

Solid Waste Management in another call for the day. Solid wastes being thrown or disposed of by households, industries and agriculture in India pose innumerable challenges which threat the environment along with its management. As per estimate, about 100,000 metric tons of municipal waste is generated daily in India. Per capita waste generation ranges from 200 gms to 600 gms per day. But the collection efficiency is 70 to 90 per cent in metro-cities and 50 per cent in smaller cities (HUDCO and HABITAT, 2001). It can be well imagined that a large portion of this generated solid waste 'must be' being contributed by industries which do not at all hold any moral obligation to move ahead for planning or managing this waste. Along with environment, other facets of human life have also got haunted with the consistent developments in science and technology. Though this form and aspect of development cannot be blamed directly for the situations of serious concerns but still, overruling the possibility absolutely will be against the analytical norms.

Agricultural Advancements and the Agrarian Crisis

Agrarian structure, in India, is called to be a primary occupational set-up, but it has also experienced lots of traumas and deterioration in the recent decades. The reason is again the 'imposed upon' conflict with industry as a consequence of technological development. Earlier, agriculture used to be the major source of national income, but now the condition, has reversed (Sridharan, 2006). The following Table shows the present condition and puts comparison of the past and present also.

Table 6.3: Inter-Sector Disparities in Income (based upon 1993-94 prices)

Year	*Agriculture*	*Industry and Tertiary*	*Ratio* $\left(\frac{Col.\ 3}{Col.\ 2}\right)$
1950-51	83154	57323	0.69
1960-61	112848	93273	0.83
1970-71	142581	153722	1.01
1980-81	167770	233392	1.39
1990-91	242012	451039	1.86
1999-2000	316780	835211	2.64

Source: Economic Survey, 2000-2001, Table 5.5, Ministry of Finance, Government of India.

The table shows that from 0.7 times in 1950-51, the disparities in income went on increasing and after 1990 galloped, this widening disparity in income acts as a pull factor for the urban area, thereby directly increasing the demand for urban land. This demand in many cases jeopardizes the country-side social and demographic set-up. Hence, no doubt the income of the nation has increased to be large extent due to industrialization, but agriculture has lost its significance as compared to the past. No doubt, agriculture sector has added consistently in the national income; still it has become, secondary to the industrial sector. Technical development has put immense economic pressure on poor farmers; made agriculture a secondary sector in spirit, thought still primary in letters; has increased debt crisis among the community of farmers; hence put multiple challenges in front of the agrarian zone of India. Excessive Technological developments as a consequence of globalization, has a direct impact on agriculture and agricultural labourers. The Green Revolution era in India, with its objectives of rural development and enquiry leading to national development, has not been replaced with pure capitalistic principles of investments, profit and business. The entry of large agricultural business firms and multi-national corporations may not promote larger national objectives and tamper with the local production structure in order to tailor agricultural production to their needs, thus generating a process of dependence of the producers on these corporations. The real loser in this shift is the small and marginal farmers and women in the Third World who would be more vulnerable to the force of the global market (Dash, 2006). A large proportion of our population is agricultural dependent. As per the 2001 census, there are about 10.70 crore agricultural labourers and more than 12 crore farmers in India. The proportion of agricultural labourers also increased from 42.6 per cent in 1993-94 to 48 per cent in 1999-2000. The World Development Report, 2002 also lists rural credit facility, urban bias public policy and dependency on the vagaries of climate as some of other problems. This is how technological development and globalization in an encapsulation, has affected the agrarian structure and communities of the nation. This was about the crisis entailed in agrarian societies due to development of science and technology. This was an attempt to justify the assertion that agrarian community has suffered a lot due to the advancements in science and technology. Though there are other illustrations to elaborate the discussion, but it may be considered as a deviation from the main theme of discussion. The gender aspect has also not been spared of the alarming impacts of scientific and technological enhancements and faces the tunes rendered thereon. The effect of scientific and medical improvements can be well reflected in the sex ratio statistics of the nation. The content is being given a laconic expression to explain the virile bias of science and technology in apposition to its environment hostile tendency.

The Endangered Sex Ratio

The 'objectivity' entailed in the structure and abstraction is squeezed into nebulous quantum, when its effect on the gender equilibrium is analyzed. Ocular gender bias is seen as an impact of scientific development. The techniques once invented to diagnose genetic disorders in unborn children, are being misused to fulfil the greed of a 'son', and, of course, science assists the expecting couple to add into the scenario. The pre-natal diagnostic techniques like ultra-sonography, amniocentesis, chorion villi biopsy, fonography, foetoscopy, maternal serum analysis, needling etc., are used to detect the gender of the child in the womb. This phenomenon proves that consciously or unconsciously, intentionally or unintentionally the frontiers of science are degrading the gender demography of the nation. These sex detection techniques (though actually meant to diagnose a disorder of any kind) certainly provoke the son preferring minds to abort a female foetus leading to a fall in sex ratio (Gupta *et al.*, 1997). Consequent to this, in some countries or regions with a strong sex preference, couples censure to have a girl child if the detection of the gender of the foetus is not an intended one (Chaudhary, 2003). A strong sex preference is interpreted as a socially determined and psychologically cradled bias of Indian society; but along with these two features, it is supported by scientific inventions also. The technological advancements and invention of novel modes of diagnosis in pre-natal situation have exhaustively influenced the reproductive behaviour in India. Sex diagnostic techniques were first introduced in mid-70s, but the motives of these inventions have been reversed drastically.

Amniocentesis was helpful to combat the risk of giving birth to a genetically deformed child, but catastrophically, these inventions were used to produce demographic adversities. The sex besotted Indian society saw an enormous potential in these techniques and hence unapologetically committed female foeticide. Though hospitals, clinics and diagnostic centres have been barred to use these techniques for sex detection, but still the trauma has not ended here. Offensive deeds have been replaced by defensive techniques, which further facilitate production of a son. The Ericsson Method separates the X and Y-chromosomes from the sperm and injects back only the Y-chromosomes into the womb, which thereby ensures a boy. The cost of this treatment is just Rs. 15,000 to 20,000 approximately. This method spares the couple to go through sex selective abortion, hence make them emancipated from any guilt too. Another method is Pre-implantation Genetic Diagnosis (PGD), which helps to determine the sex of even an unrecognizable foetus and help the couple to determine the sex of the 'to be born' child (Agarwal, 2003). The lugubrious utility instincts of Indians have in fact turned the real motives of science upside down. All such behaviours are in fact violation of the directions laid by 'Regulation for the Prevention of Misuse Act, 1994'. Though the misuse of these scientific methods to hinder the birth of a girl child has been controlled, but still, the sex deciding methods like PGD and Ericsson, will keep on emphatically contributing to the cases of female foeticide. Hence, the morose of female foeticide is assisted by scientific inventions; which further limitize the chances of improvements in the demographic profile of the nation.

A Multi-dimensional Analysis of the Impacts

It would be culpable for civic tendencies if the positivities produced by scientific and technological developments are not referred to. Certainly, human life has become more comfortable and convenient with electronic gadgets, portable communication instruments and development in research has been potent to produce panacea for many chronic diseases for which man has always been non-responding and subjugated. Many natural disasters can be predicted with the help of metrological surveys and methods and save lives of uncountable human beings. But along with these benedictions, the negativities forwarded by scientific and technological developments cannot be ruled out. The 'Millennium Project' of 'The Awareness Council' of United Nations University made an encapsulated attempt to provide a comparative analysis over the positive and negative ramification of development of science and technology. According to the report of this project, information technology possesses a potential to operationalize the term 'global', but it does intensify the impulses of fraud, cyber-terrorism, information warfare, cultural threats, etc. One of the main repercussions depicts that the low cost methods to determine the sex ratio of child can drastically affect the demographics of humanity. On the other hand, animal cloning will improve quality and cut cost of animal products, but raises ethical questions about the potential of human cloning (Ginkle *et al.*, 2000). After having an in depth gouge into the matter it is difficult to ascertain that which side of effect is more dominating i.e. positive or negative!

The common masses, especially the youth of the contemporary India, blame science and technology for an unemployed status. People hold a view that it is because of lack of knowledge of computers that they are deprived of many profiled posts and their other qualification goes into vain. Though computerization is an indicator for modernization of the economy and also of the social structure, but its addiction shapes even the careers of 'otherwise qualified' persons, into a dwarf structure. Even after all the discussions, a fact remains intact that the community reaction towards this development is also varied and changeable.

Conclusion

Development of science and technology is a transnational event which further ensures globalization because all the nations have a common aim to develop, for instance in research and science, but, if an international equity is to be achieved, the effects of this development will have to be managed. There are same putative forces which wash the iconoclast status of science as "the most objective fact" of the empirical world. If scientific and technological development is to be appreciated, the national-cum-global vulnerability will have to be dressed. In one corner of the nation, this development raises the affluence scales and at the same time, in other corner of the same nation it leads to an increase in penury. Though science is a prowess for mankind, but still it is a controlled factor which function at the 'will' of its masters, i.e. humans only. If industrialization is needed for economic development, the maintenance of environmental quality must be simultaneously stressed upon which can be done by human beings themselves. In case of female foeticide, it is difficult to determine the solutrianary paradigm, especially when population control measures have been legitimized and sanctioned. No doubt, some effects of science and technology are not controllable by human beings, but wherever it is possible, the moral, academic and physical impulses must be adjoined to control the effect of the adversities. In all a notion must always be upheld that the development of science and technology should not surpass the community sentiments and the community needs because any form of development is meant for the well-beings of individuals. Moreover to bring about human and social development in a sustainable pattern, the community itself needs to be vigilant to internalize or to reverse the impact of any event.

REFERENCES

Agarwal Anurag, *Female Foeticide—Myth and Reality*, Sterling Publishers, New Delhi, 2003.

Bharucha Erach, *Text of Environmental Studies*, University Press, University Grant Commission, New Delhi, 2005.

Briscoe, John, *Handbook of Water Persons in India*, Oxford University Press, New York, 2007.

Chari, P.R. and Gupta, Sonika, *Human Security in South Asia—Gender, Energy, Migration and Globalization*, Social Science Press, New Delhi, 2003.

Chaudhary, Manisha, *Sex Selective Abortion in Haryana*, Health Watch Trust Publication, New Delhi, 2003.

Climate Change, International Panel on Climatic Change, Cambridge University Press, London, 2001.

Dash, Satya Prakash, "Globalisation, Poverty and Agricultural Laburers", *in Sociology of Globalization—Perspectives from India* by Sakarama Somayaji Lavesh Somayaji, Rawat Publication, New Delhi, 2006.

Economic Survey, Ministry of Finance, Government of India, 2003-04.

Ginkle, *et al.*, *Human Development and the Environment—Challenges for the United Nations in the New Millennium*, Rawat Publication, New Delhi, 2003.

Gupta, *et al.*, "Abortion needs of Women in India: A Case Study of Rural Maharashtra", *in Reproduction Health Matters*, No. 9, 1997, pp. 77-86.

Housing and Urban Development Corporation (HUDCO) and United Nation Centre for Human Settlement (HABITAT), *The State of Indian Cities*, Narobai, New Delhi, 2001.

Human Development Report (UNDP), Oxford University Press, New York, 2002.

India—A Referral Annual, compiled by Research, Reference and Training Division, Publication Division, Ministry of Information and Broadcasting, Government of India, 2006.

India Infrastructure Development Report, 2006: *Urban Infrastructure*, Oxford University Press, New York, 2006.

Pant, S.K., *Human Development—Concept and Issues in the Context of Globalization*, Rawat Publishers, New Delhi, 2006.

Punjab Human Development Report, Government of Punjab, New Concept Information System Pvt. Ltd., New Delhi, 2004.

Raj Gopalan R., *Environmental Studies—From Crisis to Cure*, Oxford University Press, New Delhi, 2006.

Sridhavan, N., "Globalization and Regional Inequalities", *in Sociology of Globalization—Perspectives from India* by Sakarama Somayaji Lanesh Somayaji, Rawat Publishers, New Delhi, 2006.

UNDP, *Human Development Report*, Oxford University Press, New York, 1994.

United Nation Environment Project, *Global Environment Outlook*, Earthscan, London, 2000.

United Nation World Water Development Report, "*Water for People, Water for Life*", London, 2003.

7

Economic Policy Making and Environmental Consequences

NISHANT SHARMA and JASLEEN KEWLANI

Introduction

Humanity never denies that it fundamentally depends upon the natural systems and resources for existence and development. But in this strive to develop; it conveniently compromises the ecological concerns and cardinally aims at economic progress. Our efforts to defeat poverty and pursue sustainable development will be in vain if environmental degradation and natural resource depletion continue unabated. A comprehensive assessment of the state of the planet's ecosystems carried out by nearly 1400 leading experts from 95 countries, points out that: "The loss of services derived from ecosystems is a significant barrier to the achievement of the Development Goals to reduce poverty, hunger, and disease." Discourses and discussions of development are always undertaken in the framework of 'objectivity' as it is largely believed that all scientific events take place out of the rubrics of bias. But an abstract and even occasionally concrete contradiction can be discovered in the statement. On the one hand where development is assumed to be a magnificent symbol of the growth there only the related effects over the degrading environment have forwarded a cause of concern and revealed a serious threat to specific segment of demography. Environmentalists have long thought of economic globalization as a threat to the environment. Trade, advocated by economists get encouraged by bilateral and multilateral aid and development agencies and expanded by reduction of trade barriers both by unilateral action and reciprocal bargaining by policy-makers, is a frequent target of their anguish and anger to a larger extent, the conflict is inevitable. Impassioned differences often arise from all together different philosophies and lifestyle of trade economist and environmental activities.

The new economic policies and the structural adjustment programme often predict environmental impact without accuracy because of the complex interplay of various economic, social, political and ecological factors. These are put into effect in an effort to meet India's severe balance of payments crisis, and to propel its economy into quicker growth and global integration. A part from direct fiscal policies, the major components of the new package include boosting exports to earn foreign exchange, liberalizing industrial production, dropping barriers to the entry of foreign companies and goods expanding privatization and cutting Government spending. Economy needs to be developed for further human development, but it must not haunt the elements of "stability" in human life. Ruskin in his book titled *"Unto the Last"* also opined once that it is the object of various professions to render various things for example, a pastor's to teach, a physician to heal and a merchant (a manufacturer) to provide. It is not their object to seek a fee or a stipend. Further it is no more his function to get profit for himself; of that provision, that it is the clergyman's function to get stipend. These writings and ideas were directed at the intelligentsia which was debating the related issues, and as much they were delivered with abrasive sharpness (Mitra, 2007). This contradiction between the objects of the business makers and business-seekers still continues because it is the industrial business community, which attacks the very basic interests of the masses.

Indian companies are also squeezing natural resources at a faster pace for shrimp farming in most valuable brackish wetlands and lakes. Agri-business consortium has been proposed by the Government, ostensibly to help small and marginal farmers, which is to give a boost to commercializing Indian agriculture to cater to expanding consumerist markets. In the new liberalized atmosphere, the environment departments in all States and at the Centre are going to become everyone's punching bag and will find it harder and harder to enforce their regulations. Industries have been cleared adjacent to a Marine National Park and protected areas of wildlife officials. The destructive potential of the liberalized economy justifiably identified tropical islands, rainforests, coral reefs and marine waters and genetic store houses. World Summit for Sustainable Development in Johannesburg in its report entitled "Global Challenge Global Opportunity" has identified water and sanitation, energy, agricultural productivity, biodiversity, and human health, as the moot point. UN Environment Programme (UNEP) believed that "A consumer lifestyle based on the desire for mobility, convenience and product disposability has undercut the further advancement of resource efficiency and waste reduction". South Africa President Thabo Mbeki stated that "a global human society based on poverty for many and prosperity for a few, characterized by islands of wealth surrounded by a sea of poverty, is unsustainable. The tragic result of this was the avoidable increase in human misery and ecological degradation, as seen in the growth of the gap between North and South. The international community needs to fight the global apartheid and the elimination of inequalities between rich and poor.

Economic Development and the Environmental Concerns

The thermodynamic interpretation of the economic process therefore suggests a new definition of sustainable development which contrasts radically with present practice: sustainable development is development that minimizes resource use and the increase of global entropy. Carrying capacity can be defined as the maximum rate of resource consumption and waste discharge that can be sustained indefinitely without progressively impairing ecological productivity and integrity. Actually nearly 40 per cent of terrestrial net primary productivity (photosynthesis) is already used or co-opted by human society, one species among millions, and the fraction is steadily increasing. Hence we are within one population doubling of the 'sunshine limit' to growth and at present rates will reach that limit in 35 years (Griffin, 2006). Clearly, any human activity dependent on the consumption and use of ecological resources (forestry, fisheries, agriculture, waste disposal, urban sprawl onto agricultural land) cannot be sustained indefinitely if it uses not only the annual production of the biosphere (the 'interest') but also cuts into the standing stock (the 'capital'). Herein lays the essence of our environmental crisis. The real challenge for humanity at this millennium transition is to imagine new ways of social life, taking advantage of existing technology and knowledge to optimize the quality of living instead of maximizing the quantity of material consumption. Human species is acting as a destructive parasite on the earth, literally sacking and cannibalizing the limited earth biosphere's resources, driven to follow erroneous behaviours of economic and materialistic addiction that do not contribute anymore to the real welfare of the individuals, even less to the human society, not to mention our common life habitat: the biosphere. It becomes urgent to reimagine the available knowledge and living practices as to link humanity, in a coherent and solidarious way—to its job—to its family—to its nation—to its species—to its planet, in circular, holistic, caring, peaceful, and creative enriching evolution.

Growth has become addictive. Like heroin addiction, the habit distorts basic value judgments. Addicts of any kind are willing to pay increasing amounts for declining satisfactions. They are blind to deeper frustrations because they are absorbed in playing for always mounting stakes. Products that are new and improved promise the concept of being 'better', but leave the concept of 'whether or not good' for the individual or society completely unaddressed. Often new and better products create more wants, dependency, and dissatisfaction for most, and constantly renovate poverty for the poor (Illich, 2006; Tartajada, 2005).

In the recent decades, globalization has been so rapid and significant that it is not uncommon to see its characterization as an inexorable process. The gains to be made are so substantial for the powerful parties who drive globalization that is difficult to see the limits of this push to deeper integration. New technologies are continually being introduced which are likely to enhance these gains further (Kaplinsky, 2007). Humanity tends to usurp these developments in order to develop and grow economically. But, the strategy of consuming and digesting the development needs to be driven towards sustainability. There is a tension between the need to focus projects and interventions and the need to appreciate the complexity of the environment of the new activity. It is obvious that at any one moment, the focus may be entirely at project level with no sense of its context. Conversely, those responsible for implementations may feel they have little control over decision outside their projects (Rowlands, 2007). Simple boundary between the project and its environment is not that helpful. Smith (1981) developed a framework that recognized the environment as more complex than all the elements outside a project, or outside an organization, that cannot be controlled. As stated in the pervious parts of the paper 'poverty control' is the need of the hour along with the 'economic growth' of the nation. And to balance the two stated goals, environment inside and outside the project need to be managed.

Technology and the Environment: a Negative Correlation

Individual car, which consumes and pollutes 15 or 20 times more than train transport; the design of "modern" automobiles that emit 5 times more NO_x than 1950 models; the persistence to produce and sell disposable heavy cars of more than 1.5 litres for the use of 1 driver, 90 per cent of the time running at average speed of 15 km/h in the urban conglomerates; but as Henry Ford said, "small cars make small profits".... The result of such technological advances is that in the last 50 years about 70 per cent of all known petroleum resources have been burned: what nature created in 4 billion years, the ruling "economy" has been able to consume in 50 years! That is the real measure of the socio-economic system In-Efficiency. What we have chosen to define as progress is actually the reversal of four billion years of evolution. The money attractor and its growth imperative are calling us not forward to the future, but rather backward toward a distant past before the earth became vibrant with life.

Barry Commoner proposed to measure the environmental impact (I) as the result of multiplying three key factors: Population (P), Affluence (A), and Technology (T), so that the environmental impact of any human society may be expressed as:

$$I = P \times A \times T$$

or in words,
total pollution = population × good per capita × pollution per unit good

Source: Ehrlich, 2001.

A study of United States statistics for the period after World War II indicates that for the period from 1950 to 1967 the population increased by 30 per cent, the affluence (per capita consumption) increased by 5 per cent. In the same period beer bottles manufactured increased by 593 per cent as a result of the introduction of a throw away technology; pesticide use for crop production increased by 266 per cent, phosphate emission from detergents replacing soap increased by 1,845 per cent, NO_x from car emissions increased by 628 per cent. these disproportionate increments are only explainable understanding that, from the above mentioned three multiplying factors, the technology factor is the most influential in the resulting increment in environmental pollution, as neither the growth of the population nor the affluence factor, even combined, could account for more than 40 per cent in the environmental impact increase for the same period. The clear conclusion is that the post-war technologies were developed with the main purpose to maximize economic profit and to concentrate the economic power in new technological monopolies, without much concern for the environmental costs or the overall global social costs. It is also clear, that the environmental impact cannot be attributed chiefly to the increasing world population; therefore, the environmental problems will not be solved by simplistic "zero population growth of the others." Although population growth is a serious environmental concern, such proposals tend to mask the reality of facts to protect specific economic interests and privileges.

The environmental impact of an average inhabitant in USA is two times that of Sweden, three times that of Italy, 13 times that of Brazil, 35 times that of India, 140 times that of Bangladesh or Kenya, and 280 times that of Chad, Rwanda, Haiti or Nepal (www.cpcb.nic.in).

Before the 1973 energy crisis, that behaviour could be justified on ignorance of the limits of resources and of the boundless consumption growth consequences, or because of the cold war confrontation. Now, 25 years have passed since that time when the world community acquired a basic ecological conscience of the limits and consequences of our energetic models, and 10 years have passed after the collapse of the Soviet Union threat, but everything continues "as usual": governments and organizations as EPA, DOE, G7, IMF, WB, OECD, UN, etc., are producing tons of reports and regulations aiming to reduce up to 10 per cent the emissions, but total emissions continue to grow in proportion to the continued increase of energy consumption. In the recent Kyoto Summit (1997), the industrialized countries accepted to reduce by 5 per cent in average the greenhouse gas emission, by year 2012. As a result of the European Community pressure, but USA the number 1 world emitter may still retract from such an 'audacious agreement' if the Congress so decides. But more recently (2001), in a cynical denial attitude, the new elected USA President has rejected the signature of the treaty, by his country, the main world CO_2 emmiter, arguing that it would affect unfavourably their business economy. How miserable conception of economics! God save America. God save us from this American (USA) nightmare (Smith, 1998).

Report Analysis

Many things should be done and fast to start correcting the world economy and the world energetic technology (World Energy Projections). One measure that would combine the requirements of a sane ecology in a sane economy would be a world GDP taxation system to globally redistribute income between countries and a CO_2 emissions tax to penalize non-renewable fossil fuel consumption and stimulate the transition to a sustainable Solar Energy economy. Regarding the GDP tax, this would produce good effects reducing the level of overconsumption in the richer countries, the same as the reduction of overfeeding is a good therapy for people suffering from obesity. A reasonable GDP tax of say 12 per cent average payable by the 20 per cent richest population countries would represent a 100 per cent average income increase for the poorest 70 per cent world population. Imagine the overall economic benefit revamping the depressed economy of the 4,000 million poor people, who could access to a more decent nutrition, health services, and education; but also the benefit for the 500 million rich population in the developed countries would reduce substantially the chronic unemployment, with the huge increase in demand of basic products and services from the poor population of the world. Would not such simple common sense measures be a real and intelligent economic action, a win-win game? Where the world 'economists' are in regards to the evaluation of similar proposals? What dark forces are opposing to such possibilities?

A similar world redistribution tax on Multinational Corporation's business income would improve further the world economy, providing the required funding to help the employment casualties produced by the international corporation's business rearrangements. Would not such magnificent gestures be dignifying our extraviated human condition? After all, the planet and its resources are not a private property of any country and less of any corporation: it is the common habitat legacy for all the living species, human population included (Forbes, 2005).

The Global Warming Issue: Some Historic Factual Records and Projection in the Future

Integrated CO_2 and Temperature Information

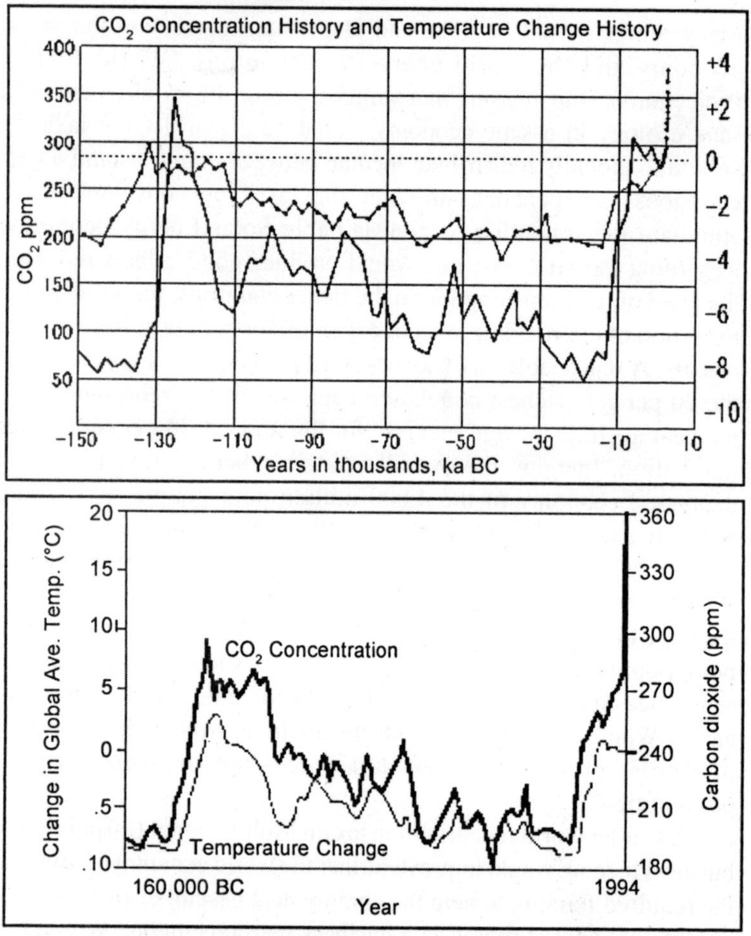

Global Temperatures and CO_2 Levels. From World Watch, 1994.

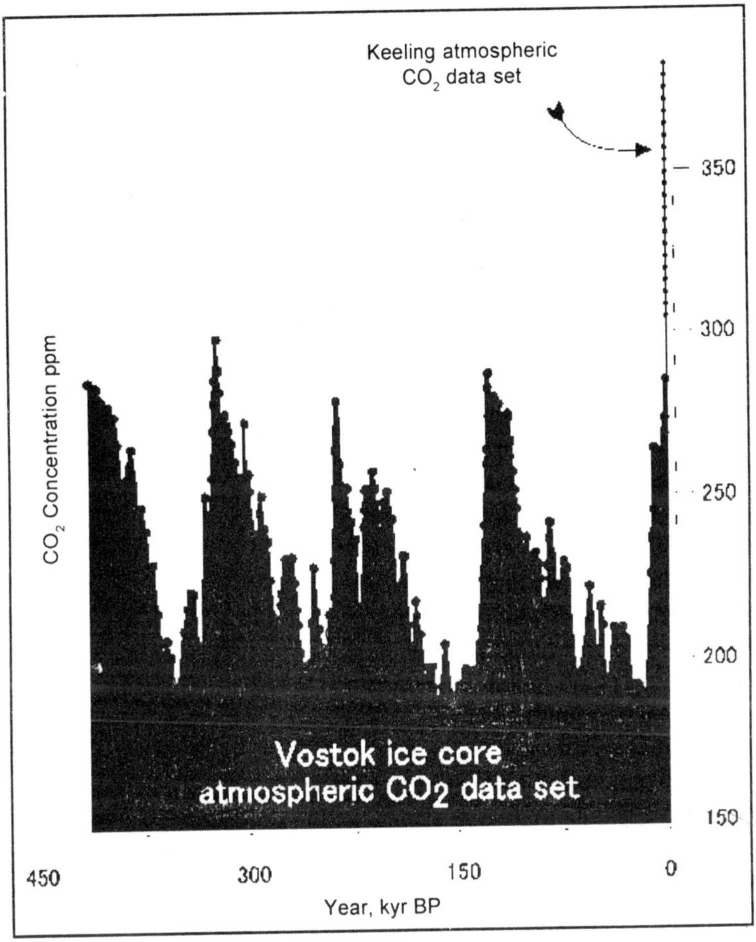

Source: Jean-Marc Barnola *et al.*, CD Keeling
Note: The source of the given three figures is the same

The graph present all known records on atmospheric CO_2 and temperature based on the Vostok Antarctic Ice Core analysis as well as the recent CO_2 atmospheric CO_2 records taken at Mauna Loa, Hawaii. Based on such record evidence, the paper presents an extrapolation of CO_2 atmospheric concentration and expected global temperature increase for a short time period from now to year 2030. In such short period there are no such factors as variations in solar radiation or other long range cosmological effects to be accounted for. What is not known exactly yet, is what atmospheric temperature increase will be necessary to produce a total melting of the polar ice. A guess value that 4°C or 6°C will cause that melting is still a matter of discussion, but that may happen by year 2020 or 2025 at the present 2 per cent rate of increase of CO_2 emissions. In the case that there is no further yearly increase, assuming that 1995 emission level are maintained constant from now on, the 4°C temperature increase will be achieved in year 2030.

The other still unknown fact is what will be the effect of temperature increase in the intensity of storms, hurricanes, flooding, changes in insect populations, extinctions of forest areas, etc, just to mention some of the possible foreseeable phenomena related with planet temperature changes. Some conservative estimates indicate that loss in property caused by atmospheric disasters has reached figures in the dozens of billions of dollars per year in the last short time period from 1998 and later, and that those losses are all time historic records. The forecast of such calamities is grim for the next years to come, and unfortunately with the actual politics of ignoring the problem and avoiding even such minimal commitments as the Kyoto protocol, things will go from bad to worse.

Here it is also suggested the close correlation between CO_2 and Temperature variation in planet history. From this, it may be inferred that recent CO_2 concentration level of 375 ppm is much higher than any value in the previous 450,000 years; and that the rate of increase of CO_2 with time is about 100 times higher than any other rate of increase in the recorded history. The greenhouse effect and the related global warming issue is a well known scientific fact established a century ago by Arrhenius, but today many 'respected' scientists in the payroll of political bureaucracies and multinational business organization, are engaging in byzantine discussion on minor technicalities, confusing and misinforming the public opinion, asking another century of further 'scientific' and academic studies to be completely sure before any meaningful corrective action is to be taken.

Environmental Sustainability

We fundamentally depend on natural systems and services such as watersheds and aquifers, coastal and marine ecosystems, forests, soils, and the atmosphere. However, these very resources are under assault from unsustainable patterns of consumption and production leading to an 'ecological footprint' that currently far exceeds the capacity of the planet. These trends disproportionately impact the poor and increasingly make them more vulnerable to natural disasters and to conflict over land and other natural resources. Unless we take decisive action, humanity everywhere will have to bear the cost in the form of degraded quality of life and growing insecurity. For example, the case of the Rainforest degradation can be refered to here.

Fortunately for all of us, the native Rainforest is practically incombustible by cause of the almost permanent rain. However, world Rainforest areas are shrinking by the combined pressure of the economic power of rich societies and the population needs of underdeveloped societies.

The Rainforest eliminates most of the gases emitted by combustion through the mechanism of photosynthesis, daily produced at huge amounts in the industrial regions of the planet and the automobiles that crowd the cities. The Rainforest regions are the ultimate and only protection available on Earth, able to dissipate the combustion emission excess of the motor driven industrial societies, so energy addicted. The Rainforests are complex ecosystems that contain and host an invaluable resource, a key factor in earth biosphere, and ultimately in the sustainability of organized life in our planet (Konda, 2006).

It is essential to impregnate ourselves in the notion that life is the consequence of an energy strange to our planet, that is solar energy. All forms of energy known to us, both the renewable (wind, hydro, etc.) as well as the non-renewable (coal, oil, gas), are or have been the result of our Sun shining over the planet. With this energy available, life on earth evolved from "simple" to "complex systems," steadily creating a life web: the biosphere. Human species is a newcomer to this long process: it required the pre-existence of this biosphere which contained all living species that preceded human. In this sense, our planet as a whole living system, may constitute a link for a common fundamental understanding of our place in Nature. From this perspective, the loss of Rainforest ecosystems is to the biosphere, as the loss of the lungs, the liver and the kidneys to a mammal as us. There is no price, no possible economic justification for such a stupid trade. The assault of human society's compulsive material consumerism is producing massive deforestation and global warming, global air and water contamination, exacerbating in large scale the delicate ecological cycles at planetary level. The so called 'technologically advanced' and overdeveloped societies are (acting as ignoring such issues and their consequences) promising that some yet unknown new technologies will correct everything when becomes necessary. But if human societies and their leaders continue to procrastinate on these issues, it will be the nature itself that will inflict the necessary corrections. Humans depend and need Nature...Nature do not need humans nor human technology. The effect of each tree being cut under deforestation is as harmful to the biodiversity, as the effect of destructing a building in a city is to human economy: loss of life and further loss of life support capacity. The global warming induced by combustion CO_2 emmisions is not a distant theoretical threat: it is a precise measurable and predictable phenomena (Mooney, 2007).

All modern economies are dependent on fixed stocks of non-renewable material and energy resources. The Second Law of Thermodynamics makes clear that they necessarily consume and degrade the very resource base which sustains them. Our material economies treat other components of the biosphere as resources, and all the products of economic activity (that is the by-products of manufacturing and the final consumer goods) are eventually returned to the biosphere as waste. Thus, while we like to think of our economics as dynamic, productive systems, the Second Law states that in thermodynamic terms, all material economic 'production' is in fact 'consumption'. Any form of economic activity dependent on material resources therefore contributes to a constant increase in global net entropy (disorder), through the continuous dissipation of available energy and matter. It follows that contrary to the assumptions of neo-classical economic theory:

— There is no equilibrium in the energy and material relationships between industrial economics and the biosphere;
— Sustainable development based on prevailing patterns of resource use is not even theoretically conceivable (Sebastian, 2007).

The following Problems and Recommendations have been identified as critical to the successful implementation of the Goals to reach a sustainable (//undp.org).

Problem 1: Integrating Environment in the Poverty Reduction Strategies

Progress in environmental sustainability depends largely on policies and programmes in sectors outside the environmental—including agriculture, forestry, fisheries, energy, transport, etc. It is, therefore, critical that Poverty Reduction Strategy Papers (PRSPs), national development plans, budgets, sectoral development plans and policies in general establish strong links between environmental sustainability and the achievement of goals.

Recommendation

Starting in 2005 we need to strengthen national and local level implementation of sustainable development strategies. Countries should make strong efforts to integrate all their environmental strategies and plans into their poverty reduction strategy. These poverty reduction strategies need to lay out direct investments in improved environmental management as well as the structural changes required to move towards environmental sustainability. Environmental measures, such water quality and protected area coverage should be used to benchmark progress towards national development goals.

Problem 2: Capacity Development for Environmental Management

Many developing countries lack the human resources and institutional capacity to formulate and implement environmental policies, strategies and plans. Furthermore, national, state and local level agencies suffer from unclear institutional definitions and frequent overlap regarding their responsibilities for environmental policy-making, management and implementation. Rather than generating a more powerful and comprehensive approach, this ambiguity often stagnates action and becomes a major obstacle to the achievement of environmental sustainability.

Furthermore the design and implementation of environmental strategies often requires highly specialized expert knowledge that is very limited or not well accessed at the country level. International organizations and the donor community do not yet provide technical advice and capacity-building support that is commensurate with local needs and at an adequate scale for implementation of national environment strategies and programmes.

Recommendation

As an important first step, countries need to:

- Strengthen their environmental institutions,
- Harmonize their legal and regulatory frameworks for environmental management, and
- Enhance environmental expertise available to the allied Ministries and local Governments—and sensitize them to the contribution of sound environmental management to poverty reduction.

Problem 3: Climate Change

The scientific evidence gathered by the Intergovernmental Panel on Climate Change and other scientific institutions is strong. Man-made climate change is a reality. It is a major development issue that is projected to have the most severe impact on countries that are least able to protect themselves and contribute very little to global greenhouse gas emissions. These are the Small Island Developing States (SIDS), coastal nations with large numbers of people living in low-lying areas, and the countries in the arid and semi-arid tropics and sub-tropics, all of which are extremely vulnerable to natural disasters and climate variability.

Despite the entry into force of the UNFCCC and its Kyoto Protocol, investments in climate change mitigation and adaptation measures remain inadequate. For example, R&D funding for renewable energy makes up a mere seven per cent of public funding for energy research. Furthermore, very few countries have actually integrated climate change considerations into national decision-making processes and development plans.

Recommendation

By 2015, the international community must agree on stabilization targets for greenhouse gas concentrations and in the near term Parties to the Convention and to the Kyoto Protocol must implement these critical agreements, particularly the industrialized world, as it is the largest contributor to greenhouse gas emissions.

Science and technology need to be mobilized now to develop the tools needed to mitigate climate change. In particular, R&D funding for renewable energy sources and energy efficiency needs to increase substantially. However, developing appropriate technologies is only part of the required response. Supporting policies and enabling environments, combined with improved mechanisms for transferring this technology, is an essential prerequisite. But more importantly, measures to promote a move to less carbon intensive options (including natural gas), less materials-intensive lifestyles and improved transport options need to be put in place. To support country-level adaptation measures, regional climate modeling needs to be further improved and adaptation strategies in areas, such as tropical agriculture, biodiversity and watershed management need to be developed and translated into workable approaches at the local level. Developing countries require the strong commitment of the international community to make the necessary transition.

Problem 4: Land Degradation

Land degradation affects as much as two-thirds of the world's agricultural land and 20 per cent of the world's pasture and rangelands have been degraded. Land degradation has become a global issue—and is no longer a local issue—as dust storms and vast landscape alteration are causing changes in regional weather patterns, including reduced rainfall. As a result, each year an estimated $42 billion in income and 6 million hectares of productive land are lost. Land degradation and unsustainable land use is leading to population migration and increasing conflicts, such as in Darfur—and dramatically increasing vulnerability to natural disasters as has been seen, for example, with Hurricane Mitch, the flooding of the Yangtze River, and elsewhere. At the heart of the problem is insecurity of land tenure, which leads to land degradation (desertification and deforestation) and poverty.

Recommendation

A major worldwide campaign is needed, spearheaded by the UN, to secure land tenure rights for farmers, herders, forest users and others. The UN also calls upon the donor community for renewed investments and sustainable financing schemes for the prevention and control of land degradation and restoration, and sustainable agricultural approaches, by scaling up the body of existing success stories around the world. Capacity development in developing countries is needed to promote innovative, community-driven technologies and solutions for integrated land management. In addition, the UN calls for increased support to the UN Convention to Combat Desertification (and its various bodies) to allow it to effectively guide Government policies.

Problem 5 : Biodiversity

The 2004 Living Planet Report measures population trends in forests, freshwater and marine ecosystems and reports declines of 40 per cent between 1970 and 2000 (www.cpcb.nic.in). This current and historically unprecedented rate of biodiversity loss—estimated to be at levels 100 to 1,000 times greater than the natural background rate—is severely undermining the critical services that biodiversity provides for the poor—and for society at large—that sustain health, livelihoods, food production, clean water, and that reduce vulnerability to impacts of natural disasters and climate change. Furthermore these environmental services are not yet valued within the prevailing macroeconomic framework. As a result, these services lack proper strategic investment and management and so are being degraded at great cost to society—although this loss does not appear in the short-term in national or global economic trend analyses or statistics. Since natural ecosystems directly provide the goods and services for the very basic existence of many of the poorest of the poor; the overexploitation of biodiversity and the loss of natural ecosystems (e.g. forests, coral reefs) is potentially the final blow that forces some of the most impoverished of the world's people to effectively become refugees.

Recommendation

Decision-makers should support the implementation of existing forest policy to curb illegal logging at local, national and global levels. The introduction of payments for ecosystem services to communities and other landowners for protection of watersheds, forest restoration and other important services, and the promotion of sustainable harvesting of natural resources through certification systems are showing great promise as instruments for stimulating local livelihoods and sound environmental management. Multilateral and bilateral organizations and NGOs should invest further in these approaches. Strengthening property and land-tenure rights, including community management regimes, and formally recognizing customary land management practices and indigenous people's rights is another vital and effective intervention, which deserves further support around the world. It is clear that trends of biodiversity loss can only be reversed if local people benefit from the conservation and sustainable use of biodiversity, in particular in countries of origin of genetic resources, and a concerted effort to scale up the wealth of successful local community experience needs to be a priority over the coming decade in order to achieve the goals. Finally, since the unsustainable use of fisheries, forests, water and land can often be traced to counterproductive economic incentives, a phasing out or redirection of environmentally harmful subsidies is needed, along with the necessary education, consensus building, provision of alternative employment opportunities and governance infrastructure.

Access to Energy

Problem 6 : Modern Energy Services

Improved access to energy, including modern cooking fuels and efficient stoves, motive power and electricity, is a critical input for reducing poverty in all its forms and promoting economic development. It is also essential to support the full inclusion of women in education and productive activities as they are disproportionately affected by heavy reliance on traditional fuels in terms of time and health burdens. Despite strong evidence for this link, existing poverty reduction strategies do not address the need to scale up access to improved energy services.

Recommendation

Countries need to scale up access to modern fuels for domestic use and electricity for productive uses. As an urgent priority, all health facilities and schools should receive access to electricity. By 2015 urban and peri-urban areas need to have access to electricity to permit the growth of a vibrant private sector and improved basic services provision. Rural areas require access to motive power at the community level to enable agricultural processing and water pumping. Countries that require external co-financing of investments in energy services should receive financial support.

Problem 7: Access to Clean Water

Water resources management is an essential ingredient to achive the desired goals—from combating poverty and hunger, ensuring environmental sustainability, improving health and reducing gender inequalities. However, some 1.1 billion people still lack access to safe drinking water. In addition, marine and freshwater ecosystems continue to be degraded due to, among others, over-fishing, land-based pollution, invasive species, dropping water tables, over-irrigation, and climate change.

Recommendation

Countries must elaborate coherent water resources development and management plans that integrate land, water and ecosystem considerations. Countries should continue to work towards developing integrated water resources management and water efficiency strategies. The development of integrated river basin schemes should be encouraged as a means to enhance sustainable ecosystem management, as well as a contribution to broader upstream-downstream dialogue.

Problem 8 : Access to Sanitation

Lack of access to sanitation is widespread and most countries in the developing world are off-track to meeting this target. Given that most of the population growth in developing countries is occurring in communities that are poor and in settlements that are informal and unplanned, the task of reaching the un-served with sanitation solutions will become increasingly difficult. Open defecation and lack of wastewater treatment represent major health and environmental hazards. Often the cause of the sanitation crisis is insufficient political attention paid by Governments and civil society to these marginal settlements where the poor live.

Substantial progress has been made in increasing access to water supply. Nevertheless, many countries are not on track to meeting this target. In many regions responsibility for the provision of water and sanitation services is increasingly devolving to communities and municipalities. Typically, however, these local level actors are not empowered—financially, technically, or institutionally to provide adequate services.

Recommendation

Increased access to water supply and improved sanitation serves several interrelated health, socio-economic and environmental objectives. Increased participation of stakeholder communities and increased investments are required in the rural sector as part of countries' poverty reduction strategies. Substantial increases in development assistance and resources need to be provided to improve the capacity of community groups and municipalities to provide water and sanitation services in urban and rural areas. In addition, the entrepreneurial capacities and resources of the private sector need to be captured in partnerships with Government to provide the necessary investments and technical expertise to meet goals in water and sanitation.

When Globalization Meets Exploitation

Trade is exploited and its virtues extolled by corporate and multinational interests, whereas environmental objectives have typically been embraced by non-profit organizations, which are generally wary of, if not hostile to, these interests. The distrust that has reflected these different traditions has led some environmentalists to extraordinary assertions of hostility toward globalization, and towards international trade and the World Trade Organization and its predecessor, the GATT. The developed world in order to maintain its hegemony through control over the resources and trading in the resources wants the underdeveloped to follow their line of action; which does not meet the objectives of globalization, rather leads to exploitation in the name of development of the developing countries. In this regard, the Fear of Japan which is emerging as a major exporter and importer is considered as rival by the United States, as discussed in the earlier chapter. Further, free trade itself is held to be a malign process, causing all manner of harm in all sorts of ways, certainly to the environment, which is rapidly becoming a key concern.

A Common Fallacy: Freer Trade, without Environmental Policy in Place, is Harmful

Thanks to the debates between free-traders and environmentalists, most sophisticated environmentalists no longer ho!d the view that if trade is freed without environmental policies being in place, not only will the environment be harmed but the country's economic welfare will be set back. But his misconception is still common in the wider environmental community. The most compelling illustration came from agricultural trade liberalization contemplated in the Uruguay Round of multilateral trade negotiations. Calculation suggest that such liberalization would shift agricultural production from higher-cost, pesticide-intensive European agriculture to lower-cost, manure-using agriculture in the poor countries, so that both income and welfare would increase in each set of countries, and total environmental quality would also improve.

Moreover, import quota protection had led, to increased imports of larger gas guzzling cars from Japan and reduced imports of smaller, higher-fuel-efficiency cars because the bigger cars carried more margin of profit than the smaller ones and it paid the Japanese car manufactures to export more of the larger cars within a given quota. So the imposition of protectionist quotas had led to both lower economic welfare and to increased pollution. But one could equally cite cases where the freeing of trade, in the absence of an appropriate environmental policy, would lead to deterioration in the environment. This is pertinently true of the import/export of agro-chemicals to the agrarian dominant economies. This seems to have been the case with what is called sometimes the "blue revolution". Rapid expansion of coastal shrimp farming in the 1980s in several countries in Asia and Latin America, principally in Ecuador, Colombia, Indonesia, Thailand, the Philippines, India Bangladesh, China, Taiwan, and Vietnam. By the early 1990s, the share of pond production of shrimp had risen to nearly one third of the total shrimp harvested.

While private domestic firms and foreign multinationals from around the world led the process, this expansion was also assisted by development specialists in the national governments and in multilateral aid agencies such as the World Bank and the Asian Development Bank. Shrimp exports constituted a growing share of export earnings, which these agencies correctly—recall assumed would also accelerate the growth of the economy. But it turned out that coastal shrimp farming, often described also as "shrimp aquaculture", was creating harmful environmental spillovers regarding which no policies were in place. Three acute problems had arisen:

- The shrimp ponds produced effluents that contaminated the water supply for others (and for themselves as well).
- They used large quantities of fresh water, which then led to a drop in the water table and the intrusion of salt water.
- Their expansion often led to the destruction of surrounding mangrove forests, "with serious consequences for commercially valuable fish and shrimp stocks which depend on a mangrove habitat during the juvenile stages of their life cycles, serious reductions in marine harvests and domestic fish supply in the future.

It will be immediately obvious in this example that if those who own and operate the shrimp farms were forced under a "polluter pay principle" to compensate those whom they harmed, then the environmental damage inflicted by the coastal farms would be reduced; for the reason is that the produces of the coastal shimps do not hold accountability so far to pay for the social cost. They do not compensate for the environmental losses forwarded through their economic activities. Economic welfare would also improve since the policy would have raised the cost of production to reflect true social costs.

The "Best" Policy: Combine Free Trade with Appropriate Environmental Policy

On the other hand, the optimal policy is to have an appropriate environmental policy in place, to look after the environment, and then to pursue free trade to reap the gains from trade. By using both policies, we get both objectives accomplished. In short, as our forefathers knew well, you cannot generally kill two birds with one stone. Those who want protection to help on the environment forget about the gains from trade that would be lost as a result of the trade protection; they are mistakenly trying to use one stone trade policy, to kill two birds, gains from trade and appropriate environment.

Income and the Environment

But the environmentalists also argue that, especially when it comes to specific pollutants and environmental harms, growing incomes will be associated with deteriorating environmental outcomes. The other argument is that global trade with and investment in countries with lower environmental standards are tantamount to "unfair competition" and will destroy our industries, which are subject to higher standards. Alternatively, they will lead, as the unions fear for labour standards, to a race to the bottom, destroying the higher standards with a view to ensuring competitiveness with lower-standard rivals. The belief that specific pollutants, such as sulphur dioxide, resulting from increased economic activity will rise in urban areas as per capita income increases depends on two assumptions: that all activities expand uniformly and the pollution per unit output in an activity will not diminish. But neither assumption is realistic.

As income rises, activities that cause more pollution may contract and those that cause less pollution may expand, so the sulphur dioxide concentration may fall instead of rise. In fact, as development occurs, economies typically shift from primary production, which is often pollution intensive, to manufactures, which are often less so, and then to traded services, which are currently even less pollution-intensive. This natural evolution itself could then reduce the pollution-intensity of income as development proceeds.

Then again, the available technology used, and technology newly invented, may become more environment-friendly over time. Both phenomena constitute an ongoing, observed process. The shift to environment-friendly technology can occur naturally as households, for example, become less poor and shift away from indoor cooking with smoke-causing coal-based fires to stoves using fuels that cause little smoke. But this shift is often a result also of environment-friendly technological innovation prompted by regulation. Thus, restrictions on allowable fuel efficiency have promoted research by the car firms to produce engines that yield more miles per gallon. But these regulations are created by increased environmental consciousness, for which the environmental groups can take credit. And the rise of these environmental groups is, in turn, associated with increased incomes.

In fact, for several pollutants, empirical studies have found a bell-shaped curve: pollution levels first rise with income but then fall with it. The economists Gene Grossman and Alan Krueger, who estimated the levels of different pollutants such as sulfur dioxide in several cities worldwide, were among the first to show this, estimating that for sulfur dioxide levels, the peak occurred in their sample at per capita incomes of $5,000-6,000. Several historical examples can also be adduced: the reduction in smog today compared to what the industrial revolution produced in European cities in the nineteenth century, and the reduced deforestation of United States compared to a century ago. Environmental activism has stimulated new regulations that have in turn produced environment-friendly improvements. All of this has made development and environmental improvement frequent bedfellows (Bhagwati, 2004).

Is the WTO Against the Environment?

But if free trade is not vulnerable to the current environmental critiques of globalizations environmentalists have also raised a different set of objections that relate not to trade as such but to its management by the World Trade Organization (WTO) and its predecessor institution, the GATT (General Agreement on Tariffs and Trade). The fact that the WTO (and the GATT, which preceded it for forty-eight years) has rules that relate to trade in products that raise safety questions (e.g., hormone-fed beef and GM products) and to trade in goods that raise values questions instead (e.g., tuna caught with nets that kill dolphins or shrimp caught without the use of turtle-excluding devices). In both cases, the environmentalists feel that the trade rules put environmental regulators at a disadvantage.

Now, generally speaking, GATT has been designed to accept whatever regulation one wants to put up, putting this within member countries' jurisdiction and exercise of autonomy. Its only requirement is that the regulation be applied equally to domestic production as to imports. Yet problems have arisen precisely because in terms of safety, values and environmental areas, regulations have been designed or operated in a fashion that has led the affected members to raise questions and even bring matters before the Dispute Settlement Mechanism.

Conclusion

Everyone in the world depends on nature and ecosystem services to provide the conditions for a decent, healthy, and secure life. Humans have made unprecedented changes to ecosystems in recent decades to meet growing demands for food, fresh water, fibre, and energy. These changes have helped to improve the lives of billions, but at the same time they weakened nature's ability to deliver other key services such as purification of air and water, protection from disasters, and the provision of medicines. Among the outstanding problems identified by this assessment are the dire state of many of the world's fish stocks; the intense vulnerability of the 2 billion people living in dry regions to the loss of ecosystem services, including water supply; and the growing threat to ecosystems from climate change and nutrient pollution. Human activities have taken the planet to the edge of a massive wave of species extinctions, further threatening our own well-being. The loss of services derived from ecosystems is a significant barrier to the achievement of the goals to reduce poverty, hunger, and disease. The pressures on ecosystems will increase globally in coming decades unless human attitudes and actions change. Measures to conserve natural resources are more likely to succeed if local communities are given ownership of them, share the benefits, and are involved in decisions. Even today's technology and knowledge can reduce considerably the human impact on ecosystems. They are unlikely to be deployed fully, however, until ecosystem services cease to be perceived as free and limitless, and their full value is taken into account. Better protection of natural assets will require coordinated efforts across all sections of Governments, businesses, and international institutions. The productivity of ecosystems depends on policy choices on investment, trade, subsidy, taxation, and regulation, among others.

REFERENCES

Bhagwati, Jagdish, *In Defense of Globalization*, Oxford University Press, New Delhi, 2004.

Commoner, Barry, *Making Peace with the Planet*, Norton Press, New York, 1992.

Cornelius, Thompson "Review" of an *Inconvenient Truth* by Al Gore 2006.

Eade, Deborah and Ernst Ligteringen, *Debating Development; NGOs and the Future*, Rawat Publications, New Delhi, 2006.

Ehrlich, P. *Wild Solutions: How Bio-diversity is Money in Bank*, Yale University Press, New Haven, 2001.

Environment and Development: 21st Century Challenges (2007) [// un.escap.org]

European Union Sustainable Development Strategy Report (2007) [// europa.eu].

Griffin, James, "OPEC, the EU and Carbon Capture and Storage" in *OPEC Bulletin*, Vol. XXXVII, No. 6, pp. 34-41. Also see "The OPEC Fund" in Vol. XXXVII, No. 1, 2006.

Illich, Robert, *Globalization and its Impacts*, Norton Press, New York, 2006.

International Union for Conservation of Nature Report (2005) [//iucn.org]

Joint United Nations Development Programme and United Nations Environment Programme Inputs on Environment Sustainability (2001) [undp.org]

Kaplinsky, Raphald, *Globalization Poverty and Inequality*, Polity Press, New Delhi, 2007.

Konda, B.R. "Impact of Globalization on Small Towns" in *Sociology of Globalization* by Sakaram Somayaji, Rawat Publications, Jaipur, 2006.

Mitra, Meera, *Its Only Business; India, Corporate Social Responsiveness in a Globalized World*, Oxford University Press, New Delhi, 2007.

Mooney, Annable, *Globalization and Uneven Development*, Routledge Publishers, Oxon, 2007.

Rowlands, Jo, *Development Methods and Approaches*, Critical Reflection Rawat Publications, New Delhi, 2006.

Sebestian, Thomas, *Globalization and Uneven Development*, Rawat Publications, Jaipur, 2007.

Smith, Niel, *The End Game of Globalization*, Routledge Press, Oxon, 2005.

Smith, W.E., The Design of Urbanization for Rural Development Projects: A Project Report, The World Bank Staff Papers, Washington, 1981.

Tartajada, Cecilia, "Critical Appraisal of Past and Present Views" in *Appraising Sustainable Development* by Asif K. Biswas and Cecilia Tartajada, Oxford University Press, New York, 2005.

8

Appraising a Local Environmental Movement in India
Some Major Determinants of Participation

DEBA PRASHAD CHATTERJEE

The central focus of this paper is analysis of a small-scale anti-pollution environmental movement in a local context. It deals with some major determinants of participation of local people in such movements in an Indian context. The present study at the same time attempts to comprehend such movements in light of Post Materialistic (PM) value change thesis as well as New Social Movement (NSM) theories. Environmental movement by now is a mature social movement and has 'become a vital part of contemporary politics in many parts of the world' (van dar Heijden, 2006: 28). Although there remain differences in opinion whether such maturity and institutionalization is a beginning of its decline or not, '...contemporary environmental movement has many of the features of an ongoing major social movement...' (Gale, 1983: 179). Variously labelled as 'ecological movement' or 'green movement' (notwithstanding some conceptual attempts to clarify the subtle differences in this regard), environmental movement is delineated as '...a (very loose) non-institutionalised network which includes, as well as individuals and groups who have no organizational affiliation, organizations of varying degrees of formality and even parties, especially Green parties, and which is engaged in collective action motivated by shared environmental concern, but that the forms and intensity of both action and concern may vary considerably from place to place and from time to time' (Rootes, 1997a; see also Rootes, 1997b). For environmental sociologists, environmental movement represents perhaps the most important form of collective behaviour to be reckoned with at this point in time. As an important field of research, environmental movement '...predated the institutionalization of environmental sociology and has remained important through the mid-1980s' (Buttel, 1987: 475).

Mainstream environmental movement literature in India stresses on much-publicised ecological conflicts surrounding livelihood issues, thanks to the universal role of mass-media (Chapman *et al.,* 1997; Hansen, 1993; Gadgil and Guha, 1994; Yearley, 1991) in pointing out environmental issues to public eyes. As Shiva (1986: 256) affirms, 'Ecology movements in India are struggles of the dispossessed, the marginalized, the victims of discrimination, among whom can be counted women, tribals and the non-commercial farmers. These movements are aimed at conserving nature's balance in order to conserve their means of survival.'

Authors generally consider Indian environmental movements by dividing them into some broad categories such as, *Forest-based struggles* (Chipko, Appiko, for example), *Struggles over land use* and *anti-SEZ movements* (struggle in Tawa Command area of Madhya Pradesh, struggle of tribal people against the Bharat Aluminium Company—BALCO in the Gandhamardan Hills of Orissa, anti-mining movement of Doon valley in the Himalayas, and more recently, against establishment of the Special Economic Zones (SEZ) in different parts of the country as Orissa, Goa, West Bengal for example), *Struggles against big dams* (Narmada Bachao Andolan—NBA, Tehri, Silent Valley, Koel Karo and Icha dam movements besides others, for example), the *Battle over the seas and marine resources* (such as Kerala fish-workers' struggle against large-scale commercial fishing. These are some of the major and widely known areas of environmental activism in modern India (See, Sethi, 1993; see also Shiva, 1986; 1988; 1991; 1992; 1994; Guha, 1989; Singh, 2001). Besides all these much-publicized people's resistance events there exists a relatively feeble force of environmental protests at the community levels, often in small urban localities that Sethi (1993: 134) terms as the *Struggles against pollution.*

Differently labelled as environmental or ecological movements of various strands, authors sometimes question the very appropriateness of conferring such labels on these movements.

Commenting on the trend-setting achievement of courageous Chipko activists in an article on the 25th anniversary of the movement, Bandyopadhyay (1999: 880) argues, 'The movement had its beginning in the conflicts over mountain forests between the economic interests of the mountain communities and the economies of the plains'. Subsequent developments raise several questions about the environmental or ecological nature of the movement. According to Ramachandra Guha (1992: 224), '...Chipko is best viewed as a peasant movement.' In line with the distinction made by Guha (1989), Bandyopadhyay (1999: 881) further states, '...the Chipko movement has the private face of a quintessential peasant movement and a public face of one of the most celebrated environmental movements of the world.

Chipko has its roots in the hard economic struggle for survival, while its face has been tacitly decorated by some "deep ecological" terms. Insufficient attention paid to the political and economic dimensions of the Chipko protests is also highlighted by the scholars. At the same time ecofeminist and ecocentric portrayals of the aforesaid movements besides the iconic 'Chipko' have also been criticised theoretically as well as empirically (Jackson, 1993). Similarly, Mawdsley (1998: 50) rejects the neo-populist interpretation of the movement and comments, 'Chipko developed primarily as an economic struggle...within the geographical constraints of the mountains. Environmental concern was not lacking, but a sound environment was seen as a *functional requirement for a sound local economy*. Within this, different "subsistence strategies" were articulated at different times, in different places, and by different people...' In view of the aforesaid criticisms, one can consider the Chipko movement as a part of the long drawn-out protest movement of the disadvantaged hill people against the multi-faceted deprivations of the State, rarely as an environmental movement *per se*. Subsequent political mobilisation in the Uttarakhand region and formation of Uttaranchal State, notwithstanding how far the new State would be able to address the problems of the hill people (See Jayal, 2000), can strengthen such an argument.

Why is there such an inclination on the part of the scholars as well as the mass-media to publicise primarily subsistence movements of the disadvantaged lot of Indian society as ecological or environmental movements? For Chipko, besides the logic of naïve populism (Mawdsley, 1998), Bandyopadhyay (1999) also points to the media created confusion by the urban environmentalists. Nevertheless, a closer look reveals some other possible explanation too.

First, the retreat of class analysis and the emergence of post modernist project throughout the 1970s and the 1980s probably guided the scholars as well as the upbeat mass-media (consequent to the 'fall' of 'communism' in the USSR as well as East European countries) to construct some non-class analysis of the social movements of the disadvantaged and marginalized people at that period. Consequently, Chipko, the subsistence movement of the deprived hill people has been labelled as an environmental movement. In the West, relatively affluent/middle class public-led environmental movements could define themselves without reference to class. This is difficult and seems rather non-realistic for underprivileged India where class analysis ought to be more plausible instead of taking up a post modernist rhetoric as environmentalism. Nevertheless, scholars, mostly Indian as well as a few Western, had to take recourse of post modern environmental paradigm possibly in view of the retreat of class analysis, especially during the 1980s. The periodicity of the Chipko movement (in the early 1970s); its labelling as ecological/ ecofeminist or environmental and glut of literature primarily in that direction (in the 1980s and early 1990s; in fact an earlier treatise bears no mention of Chipko at all; see Dasgupta, 1978) and subsequent criticism and rejection of the predominant literature (mainly in the late 1990s), points to the validity of such an explanation.

Second, it is possible that rise of modern environmentalism and consequent proliferation of social science literature on environmental issues in the last quarter of the 20th century, primarily in the Western context, drew attention of the Indian scholars into that orbit. Western influence on Indian knowledge industry from the onset of modern age is now a well-accepted fact. Apropos his concept of 'Westernization', Srinivas mentions about a group cf Indians who '...absorbed Western science, knowledge, and literature...' (Srinivas, 1962: 51). For D'Souza (1998: 11), 'Indian sociologists ...have had to depend upon the theoretical knowledge developed in the West for interpreting social phenomena in their society'. He further argues, '...some of them concluded that the Indian society was not comparable with the West and so it needed altogether a different kind of conceptual framework and theory. This was also the opinion of some of the British and European anthropologists.' Such Western influence, perhaps, has remained instrumental in marking some of the newly emerging mobilizations of the dispossessed Indian people centring primarily the livelihood issues, as environmental movements, possibly to draw the attention of the dominating Western trend in particular, and the world community, in general.

Third, political dimension of such relatively large-scale grassroots movements of the dispossessed also needs to be explored. Some of the local political elites probably found these movements handy to strengthen their regional demands for varying degrees of power, ranging from local autonomous councils to separate Statehood. As Mawdsley (1998: 43) argues, 'Despite their many differences, the Chipko protests and the regional movement clearly had/have their origins in similar or overlapping issues. However, these contiguities are *not* apparent from a reading of many neo-populist accounts of the Chipko protest(s), which offer a very different set of understanding about the environment, development and often gender relations in the hills.' For Mawdsley, the Chipko movement represents a variant of hill peoples' protest against internal colonialism by the State of Uttar Pradesh. Movement of tribal peoples of Chotanagpur region against Subarnarekha multipurpose river valley project and the construction of Koel Karo and Icha dam can also be marked as a form of protest against the internal colonialism by a developmental State, rather than being an instance of environmental movement. Terming these collective mobilizations as 'environmental' or 'ecological', as it seems, could only serve the local political elites' propaganda interest in a better way, as the prefix 'environmental' could be a magnet for better mass-media attention throughout the 1980s and the 1990s. Formation of at least two new Indian States in the year 2000 (GOI, New Delhi) i.e. Jharkhand and Uttaranchal, surrounding the hotbeds of such movements and subsequent weakening of the movement forces after State formation possibly explain the motive of such labelling of these movements as environmental/ecological.

In short, these protest movements of the dispossessed mass of India for several reasons have been labelled as environmental/ ecological movements in general by media as well as the scholars during the last two decades of the 20th century and such a tendency is still continuing. At the same time due to relative infancy of the institutional development of environmental sociology in India, relevant sociological intervention in this domain of social movements is also scanty. Altogether, it seems that the startling growth of environmental sociology in Europe (Foster, 2002) during the 1990s, with its characteristic philosophical sophistication besides an equal under emphasis on empiricism/naturalism/realism of the American tradition, is yet to take off in India. This may possibly explain the relative absence of environmental sociological literature on environmental movements in India. In consequence, relatively small-scale, localized, often sporadic grassroots anti-pollution environmental/ecological protests surrounding the quality of life issues, primarily confined to urban localities, have remained relatively unattended. This is despite recent salience of industrial pollution in social science literature (Yearley, 2005). In many a sense, this is an important gap in our knowledge. The study at hand humbly strives to bridge the gap in its attempt to appraise a small-scale anti-pollution environmental movement in a local context.

A few conceptual clarifications seem necessary at this stage. First, 'environmental movement', a key phrase of the present study, as distinct from ecological movement is often used by scholars synonymously. Sethi (1993: 128) mentions about various types of struggles and conflicts over the control and use of natural resources and differentiates between environmental and ecological movements. For him, movements characterised by 'anti-pollution and conservationist consciousness' should be termed as 'environmental' in contrast to 'ecological' movements that hover around more fundamental issues of rejecting 'the dominant developmental paradigm and seeks to alter the very classification of both man and nature on which notions such as progress are based.' While such a differentiation might evolve serious debates, for the purpose of the present study, such a conception of environmental movement seems fairly satisfactory.

Second, for all purposes in the present study, *pollution* stands for the presence of any unwelcome substance in the environment that is detrimental to health and well-being of any form of life on the earth—human as well as non-human beings.

Third is the concept of 'local'. Scholars of diverse streams variously use the term while its meaning varies from a small neighbourhood of, say, about fifty persons to a large nation of fifty hundred thousand people. Many a time, instead of such a numerical dimension, 'local' is viewed in terms of different cultural traits such as language, ethnic identity, religion, political boundary and so on. In fact, 'local' refers to space that has been given meaning through, individual, group or cultural processes. As Vorkinn and Riese (2001: 252) mentions about 'place attachment', 'local' too possibly involves 'an interplay of emotions, cognition, and behaviour, and it is not a static phenomenon but may vary over time.' Again, Yanitsky (1999: 159) defines 'local context' as '...the immediate economic, political and social milieu...' Rannikko (1996: 58) indicates 'local' as 'a specific physical environment'. For the present purpose 'local level' stands for 'community level' as a specific physical environment characterized by the immediate economic, political and social milieu of an individual.

Regarding the salience of environmental movements at the local level, Buttel (1987: 476) notes an '... increased emphasis on local or specialized environmental groups such as the antinuclear and appropriate technology movements, as well as local movements generated by toxic waste and related environmental problems.' More recently, Rootes (2007a and 2007b) points to the ubiquity and recurrence of local environmental movements, even in times when environmental issues are not salient on national agenda. Yet, their relationship to trans-local environmental movements and issues has been relatively neglected. He further argues that the relationship between local campaigns and global environmental issues is problematic, but the ways in which local mobilisations often combine issues of environment, economic justice and democracy mirror the emerging agenda of transnational environmentalism.

On the spurt of environmental problems in industrially developed societies and people's protest against such problems during 1970s, Habermas (1981: 35) comments on the global-local relevance as following:

> '... The large industrial intervention in ecological balances, the scarcity of non-renewable natural resources, and the demographic development present industrially developed societies with serious problems. Yet these challenges are largely abstract and require technical and economic solutions that must, in turn, be planned

> globally and implemented by administrative means. What sparks the protest, however, is the tangible destruction of the urban environment, the destruction of the countryside by bad residential planning, industrialization and pollution... These are developments that visibly attack the *organic foundations of the life-world* and make one drastically conscious of criteria of liveability, of inflexible limits to the deprivation of sensual-aesthetic background needs.'

Habermas' emphasis on the problem of pollution that sparks protest in the industrially advanced countries notwithstanding, rapidly industrializing countries of the global South are increasingly facing a similar situation too. CSE (1999: 281) quotes a World Bank study in reporting that the largest share of economic and health costs resulting from environmental damage is due to growing water and air pollution and premature deaths due to air pollution is estimated to have increased by thirty per cent during 1991-1995 in the Indian cities. It also reports that poor suffer the most from environmental damage and illness, as polluted water and air proves to be far more costly to the poor than the rich. In a recent study that attempts to compare resource use and industrial pollution cases across countries with varying political systems, levels of social and economic development and complexes of cultural values that also includes the case of Delhi urban air pollution, Yearley (2005) argues that the definition of pollution and of the most important kinds of pollution can be subject to discordant interpretations within a single culture (notably in case of Delhi, between middle-class and trade-union representatives). Again, developmental intentions of the state lead to the formulation of non-stringent anti-pollution laws in India and the problem of effective implementation of legal measures also remain there. Such a backdrop ostensibly offers a lush ground for protest movements in different parts of India.

While such movements might be emerging too at the local level, in the absence of any large-scale pan Indian movement organization (except The National Alliance of People's Movement with untiring Indian environmental crusader Medha Patkar as its coordinator) and relative dearth of mass-media attention, such movements often might remain unreported. This is not to deny the validity of the underprivileged people's aspiration for higher quality of life-induced-apathy to initiate or take part in any environmental movement that can jeopardise their growth interest; as such movement is often perceived as 'anti-developmental'. On this point, Sethi (1993: 134) comments, 'Struggles against pollution *per se* have not been too popular in India.' Without necessary empirical substantiation, he further argues that such struggles are '...confined to complaints by the better-off sections in our cities about the quality of the air and water they consume, resulting in some state action of both a regulatory and a provisionary kind.' (*ibid.)*

Nevertheless, such movements are coming up in India. In the same article, Sethi mentions about several collective mobilizations against fresh water as well as river water pollution as a new trend. He also mentions about some protests of the rural farmers against the pollution caused by agro-based industry and allied chemical factories (*ibid.*: 135). CSE (1999) reports about a plethora of public interest litigations in the Supreme Court of India (the apex court of the country) related to the people's protest against the problem of pollution where the judges have shown their capacity to '...make legal innovations to address the growing environmental crisis in the country' (CSE, 1999: 360). The agitation of the villagers against the Coca Cola bottling plant responsible for over-exploitation of the ground water resources and pollution of sources of potable water in Plachimada village (Venugopal, 2003: 87), Kerala represents a recent example of such anti-pollution mobilizations at the local level. A large number of such efforts often remain unmarked too.

Present study tries to analyse such a relatively unnoticed environmental protest movement at the local level, as a social movement. During late 1980s and early 1990s, people of Chandrahati village and Tribeni Township, district Hooghly, West Bengal, initiated a protest movement against the state owned Bandel Thermal Power Station (BTPS), and two other privately owned chemical factories, the TTL Limited and the Kesoram Rayon (KR). Their protest was against the menace of flyash of BTPS, chemical pollution of local fresh water, land and river ecosystem for which they believed TTL and KR as responsible. With the charge of severe air pollution on its shoulder, the BTPS, a public sector company, owned by the Government of West Bengal, topped the list of polluting industries in the area.

The protest movement, led by the Chhai Pirito Nagarik Committee (CPNC), primarily located in the Chandrahati and Benipur Villages besides Tribeni township, could also yield some respite for the local inhabitants as the authorities of the factories, of late, have taken some measures towards solution of the problem. The movement, led by the local people of different strata, perhaps first of its kind in this area, got discrete local and regional media attention to some extent at that time. Besides regular protest deputation to the concerned authorities and legal routes, the leadership also adopted several innovative forms of collective mobilisation against the 'negligent' executives. Expectedly, such a movement raised the environmental concern of the local people and they also encouraged the present researcher to undertake a social scientific study of the problem and co-operated all the way. Although the movement is a feeble force now, its optimist leadership is still watchful and, as some of them believe, can generate fresh waves of mobilisation, whenever required.

The movement seems analysable in light of post materialistic (PM) value change model and New Social Movement (NSM) theories. In this respect, the available literature is ambivalent leading to an unresolved ambiguity about applicability of PM and NSM models in the context of developing countries. The conventional wisdom represented by Inglehart (1971, 1977, 1983) suggests that residents of the poor, non-industrialised nations rarely exhibit PM values leading to inapplicability of NSM model in analysing recent social movements in these nations (see also Buttel, 1987, 1992, Dunlap and Mertig, 1995 and Brechin and Kempton, 1994 for a few additional citations and Franzen, 2003, Chatterjee, 2008 and Gillham, 2008 for more recent findings in support of the conventional wisdom). Conversely, available evidences suggest just the reverse as equally valid (Dunlap and Mertig, 1995; Dunlap *et al.*, 1993; Brechin and Kempton, 1994; Lundy, 1999; Gadgil and Guha, 1994). In the present study, as the majority of the local residents are employed in the three factories, they also expectedly represent a lower, lower-middle or middle class background. Usually environmentalism is seen as a middle class phenomenon. Emergence of a sizeable middle class in India during the last decade (Beteille, 2002) demands increasing attention of the sociologists dealing with Indian society towards quality of life issues rather than their erstwhile importance on the subsistence issues. As anti-pollution environmental movements represent such issues, sociological deliberations on such movements in light of the PM value change model and NSM theories assume growing salience. The study at hand is focused towards that end.

Studies on small scale, localised anti-pollution movement, particularly from a sociological perspective, are nearly non-existent in India. On the other hand, a good number of contributions go to the credit of the scholars representing a diverse social scientific background (Desai, 1985; Karan, 1980, 1994; Kumra, 1982). Although environmental issues are gradually receiving attention of Indian sociologists of late (Bapat, 1988; Jathol, 1991; Baviskar, 1995, 1997a, 1997b; Chatterjee, 1998-99, 2004, 2008; Munshi, 2000; Oommen, 1990) a general indifference in taking up empirical research is also notable here. At the same time, in the context of growing environmental awareness of the Indian people lead by the state (especially noteworthy is the increasing intervention by a proactive judiciary; see CSE, 1999), the NGOs and the mass media in recent years, studies on environmental activism deserve special importance from the students and practising sociologists. Present study strives to fulfil this gap in knowledge.

Objectives

Exploration of the extent to which different socio-demographic variables determine participation in environmental movement is an important objective of this essay. For participation in any environmental movement, level of environmental concern of an individual is generally considered as a precondition and such concern is often associated with her objective experience of pollution, besides a host of socio-demographic variables such as age, education, family income, residence etc. The primary objective of the present study is to understand whether and to what extent existence of pollution is a significant factor that can differentiate people's participation in the CPNC movement—a local level environmental protest movement. At the same time, it also attempts to explore the complex interaction of different socio-demographic variables in determining people's commitment to participate in such movements. The study also makes an effort to understand and analyse the movement in the light of PM value change and NSM theories with the help of an in depth interview of the leadership of the movement.

Methodology

A household survey of people having practical experience of the problem of pollution is carried out with a semi-structured questionnaire. For the survey, altogether 375 respondents have been interviewed with a semi-structured interview schedule that contains 138 items/questions. The schedule consists of four portions. Personal identity of the households is covered in the *first* part. Issues related to general socio-economic background of the respondents are covered in the *second* part. In the *third* section, items reflecting respondents' knowledge and awareness of the problem of pollution are included. Finally, in the *fourth* segment, questions on respondents' commitment to act for abatement of the menace of pollution and their willingness to participate in any movement for better environment are placed.

Some selected items from the fourth part of the schedule reflecting an individual's level of participation in the CPNC movement are the primary *dependent variables* in this study. These are about respondent's willingness to participate in environmental movement; local as well as relatively more general and larger scale. Degree of participation is specifically reflected in terms of the amount of time and money an individual is willing to devote for the cause. Residence (in high or low pollution zone) and different socio-demographic variables such as age, education and family income are the major *independent variables*.

The 375 households visited for the purpose of survey are located in two hypothetical concentric zones. The High Pollution Zone (HP) consists of 189 sample households and the remaining 186 are from the Low Pollution Zone (LP). Items related to actual participation in the CPNC movement are only meaningful for the 149 respondents of Chandrahati, Benipur and Tribeni, the three HP localities where the movement remained largely confined. These are primarily related to a respondent's knowledge of the existence of the movement and its contemporary status, information regarding participation, reasons for non-participation, and mode of participation in the movement besides her views about its success or failure.

A separate questionnaire is administered among the top brass of the movement. It includes items pertaining to actors' socio-demographic background information, their opinion about such background of the support base, structure of the movement location, goals or ideology, organization or structure, medium of change/ tactics, participation etc. that differentiate conventional social movements from the newer ones (Martell, 1994: 112, see also Pichardo, 1997: 414). Twenty-four such movement leaders have been selected for this purpose.

Findings

Participation in the environmental movement is sought to be understood in the present paper on the basis of three items; first—expressed willingness of the respondents to participate in any environmental movement; local as well as general or large-scale. In the more focused second and third items, their willingness to devote time and money both for any local as well as general environmental movement has been explored. The extent to which level of participation varies on account of different socio-demographic variables such as age, education and income of the respondents is classified, both for High and Low Pollution zones.

Table 8.1 shows zonal variation in expressed willingness to participate by age of the respondents. The modal category of response, as the table reflects, is the high willingness to participate in any environmental movements, local or general, irrespective of age, both in the High and Low Pollution zone. At first glance, age fails to determine participation in environmental movement.

A closer look, nevertheless, reveals that the respondents of higher age groups in both the zones are highly interested to participate in local environmental movements compared to general, and this is markedly greater than the younger ones, particularly in the HP. This indicates failure of age to determine willingness to participate in the CPNC movement; contrary to conventional wisdom (Michelson, 1976; Van Liere and Dunlap, 1980; Buttel, 1979; Buttel and Flinn, 1976; Hamilton, 1985; Jones and Dunlap, 1992; Lowe *et al.,* 1980) that suggests a dominance of youths in this regard. Nevertheless, findings here points to the importance of zonal difference in such willingness to a greater extent as respondents of both the age categories in High Pollution zone exhibit greater willingness to participate in any environmental movement, general or local, than their counterparts in the Low zone. Thus, objective existence of pollution in one's locality emerges as an important factor, as some studies report, in influencing her willingness to participate in environmental movement in contrast to age as a socio-demographic variable (Martell, 1994: 132; Tester, 1991; Lowe and Rudig, 1986; Jehlicka, 1992; Jehlicka and Kostelecky, 1991; see also, Beck, 1992 and Giddens, 1990; 1991).

Table 8.1: Zonal Variation in Willingness to Participate in Environmental Movement: General *versus* Local by Age Groups

Age	*Willingness to Participate*								*Total*
	Unwilling		*Undecided*		*Willing*		*Very much willing*		
	General	*Local*	*General*	*Local*	*General*	*Local*	*General*	*Local*	*N*
High Pollution Zone									
Lower (Up to 40 years)	8.43	**10.8**	13.25	**9.6**	31.33	**15.66**	46.99	**63.86**	83
Higher (51 years and above)	11.9	**10.2**	23.73	**8.5**	32.2	**15.25**	32.2	**66.1**	59
Total	9.86	**10.6**	17.61	**9.2**	31.69	**15.49**	40.85	**64.79**	142
Low Pollution Zone									
Lower (Up to 40 years)	18.9	**5.26**	17.89	**15.8**	35.79	**25.26**	27.37	**53.68**	95
Higher (51 years and above)	29.8	**25.5**	8.51	**8.51**	44.68	**25.53**	17.02	**40.43**	47
Total	22.5	**12**	14.79	**13.4**	38.73	**25.35**	23.94	**49.3**	142

Table 8.2 indicates more specifically about the age differential of respondents' willingness in terms of their willingness to devote time for environmental movement, local *versus* general. Respondents who desire to devote less than hour a day or holiday and those who wish to spend 2 to 3 hours or more a day/holiday are categorised as 'Willing' and 'Very much willing' respectively, for the purpose of the present study. Instead of 'Very much willing', here the modal category is 'Willing', indicating a notable departure from their earlier position about 'Willingness to Participate' *per se* as depicted in Table 8.1. Here too, the bias towards local environmental movement is observable generally in both the age categories, except among the respondents of higher age category in High Pollution zone, who

exhibit their lesser inclination to be 'Very much willing' to devote time for local environmental movements compared to general.

Table 8.2: Zonal Variation in Willingness to Devote Time for Environmental Movement: General *versus* Local by Age Groups

Age	*Willingness to Devote Time*								*Total*
	Unwilling		*Undecided*		*Willing*		*Very much willing*		
	General	*Local*	*General*	*Local*	*General*	*Local*	*General*	*Local*	*N*
High Pollution Zone									
Lower (Up to 40 years)	22.9	**9.64**	30.12	**21.69**	44.58	**62.65**	2.41	**6.02**	83
Higher (51 years and above)	20.3	**8.47**	42.37	**37.29**	32.2	**50.85**	5.09	**3.39**	59
Total	21.8	**9.15**	35.21	**28.17**	39.44	**57.75**	3.52	**4.93**	142
Low Pollution Zone									
Lower (Up to 40 years)	29.5	**13.7**	25.26	**21.05**	41.05	**60**	4.21	**5.26**	95
Higher (51 years and above)	42.6	**25.5**	31.92	**25.53**	21.28	**44.68**	4.26	**4.26**	47
Total	33.8	**17.6**	27.47	**22.54**	34.51	**54.93**	4.23	**4.93**	142

A good proportion of respondents of both the pollution zones are by and large unwilling to give over time in general/large-scale environmental movement compared to local.

However, proportion of older age category respondents unwilling to offer time for local as well as general movement is also noticeably high in low pollution zone compared to high, signifying the importance of existence of pollution in their physical as a significant determining factor here too.

Table 8.3 reports respondents' willingness to contribute money for any environmental movement. 'Willing' stands for their willingness to contribute less than Rs. 10 per month whereas 'Very much willing' stands for their willingness to contribute more than Rs. 20 per month for the purpose. While local bias is observable in general, younger respondents differ considerably in the 'Very much willing' category as they show higher inclination for the local movement than their older counterparts in support of conventional wisdom. A comparison between Table 8.2 and Table 8.3 in the 'Willing' category as the modal category reveals greater proportion of respondents prefers to donate a little amount of money than devoting a little amount of time, irrespective of their age, for the environmental cause.

Table 8.3: Zonal Variation in Willingness to Contribute Money for Environmental Movement: General *versus* Local by Age Groups

Age	*Willingness to Contribute Money*								*Total*
	Unwilling		*Undecided*		*Willing*		*Very much willing*		
	General	*Local*	*General*	*Local*	*General*	*Local*	*General*	*Local*	*N*
High Pollution Zone									
Lower (Up to 40 years)	20.5	**4.82**	21.69	**19.28**	54.22	**60.24**	3.61	**15.66**	83
Higher (51 years and above)	11.9	**6.78**	25.42	**16.95**	52.54	**59.32**	10.17	**16.95**	59
Total	16.9	**5.63**	23.24	**18.31**	53.52	**59.86**	6.34	**16.2**	142
Low Pollution Zone									
Lower (Up to 40 years)	18.9	**6.32**	15.79	**8.42**	53.68	**68.42**	11.58	**16.84**	95
Higher (51 years and above)	23.4	**17**	17.02	**12.77**	53.19	**59.57**	6.38	**10.64**	47
Total	20.4	**9.86**	16.2	**9.86**	53.52	**65.49**	9.86	**14.79**	142

Table 8.4 shows inter-zone variation in expressed willingness to participate by educational levels of the respondents. 'Very much willing' once again emerges as the modal category of response. Once again, bias towards participation in local movements is clearly evident. A marked difference on the basis of educational level is also clear to notice. Whereas in High Pollution zone proportions of lower educated respondents very much willing to participate in local and general environmental movements are 58.33 and 35.19 per cent respectively, such proportions for higher educated respondents are 69.39 and 48.98 in that order. Similar trend is also noticeable for respondents of Low Pollution zone signifying role of education as an important determinant of participation in environmental activism in support of conventional wisdom (Buttel and Flinn, 1974; Kumra, 1982 see also Van Liere and Dunlap, 1980).

Table 8.4: Zonal Variation in Willingness to Participate in Environmental Movement: General *versus* Local by Educational Levels

Education	*Willingness to Participate*								*Total*
	Unwilling		*Undecided*		*Willing*		*Very much willing*		
	General	*Local*	*General*	*Local*	*General*	*Local*	*General*	*Local*	*N*
High Pollution Zone									
Lower (Up to Secondary)	13	**15.74**	21.3	**11.11**	30.56	**14.81**	35.19	**58.33**	108
Higher (Graduate and above)	12.2	**4.08**	10.2	**12.24**	28.57	**14.29**	48.98	**69.39**	49
Total	12.7	**12.1**	17.83	**11.46**	29.94	**14.65**	39.49	**61.78**	157
Low Pollution Zone									
Lower (Up to Secondary)	30.2	**13.21**	11.32	**13.21**	33.02	**28.3**	25.47	**45.28**	106
Higher (Graduate and above)	9.52	**6.35**	20.63	**12.7**	42.86	**23.81**	26.98	**57.14**	63
Total	22.5	**10.65**	14.79	**13.02**	36.69	**26.63**	26.04	**49.7**	169

Table 8.5 indicates more specifically about the educational status differential of respondents' willingness in terms of their willingness to allocate time for environmental movement, local *versus* general. Here again 'Willing' emerges as the modal category and predisposition towards Local is clear.

However, education perhaps fails to determine willingness to devote time, as respondents of lower educational level show as much enthusiasm for environmental activism as their counterparts of higher educational level show; both in the High as well as the Low Pollution zones. Thus, a sizeable proportion of respondents with higher educational level, despite their willingness to participate in environmental movement, as reflected in Table 8.4, are not agreeable to devote time for the purpose. Such a finding stands against the conventional wisdom and indicates that in matters of down to earth routine activism, educated people hesitate to some extent. This also challenges the charge of elitist bias that often marks modern environmentalism and supports studies in that order (Morrison and Dunlap, 1986; Brechin and Kempton, 1994; Dunlap and Mertig, 1995). On the other hand, greater enthusiasm of lower educated, generally disadvantaged people to devote time can also be related to their relative inability to support the movement financially. Another possible explanation lies in their general inclination to support any ameliorative activism against the established order, in this instance, the environmental movement.

Table 8.5: Zonal Variation in Willingness to Devote Time for Environmental Movement: General *versus* Local by Educational Levels

Education	*Willingness to Devote Time*								*Total*
	Unwilling		*Undecided*		*Willing*		*Very much willing*		
	General	*Local*	*General*	*Local*	*General*	*Local*	*General*	*Local*	*N*
High Pollution Zone									
Lower (Up to Secondary)	27.8	**12.96**	31.48	**25**	35.19	**58.33**	5.56	**3.7**	108
Higher (Graduate and above)	20.4	**10.2**	30.61	**28.57**	42.86	**51.02**	6.12	**10.2**	49
Total	25.5	**12.1**	31.21	**26.11**	37.58	**56.05**	5.73	**5.73**	157
Low Pollution Zone									
Lower (Up to Secondary)	42.5	**19.81**	19.81	**15.09**	33.02	**56.6**	4.72	**8.49**	106
Higher (Graduate and above)	14.3	**7.94**	42.86	**33.33**	38.1	**55.56**	4.76	**3.18**	63
Total	32	**15.38**	28.4	**21.89**	34.91	**56.21**	4.73	**6.51**	169

As observed in connection with age differential of willingness to put in money, Table 8.6 also exhibits 'Willing' as the modal category. Both in High and Low Pollution zones respondents are more 'Willing' or 'Very much willing' to support local environmental movements compared to general and a considerable edge of higher educated respondents over their counterparts in contributing money for environmental movement is also evident from the 'Very much willing' category, in support of conventional wisdom.

As the table shows, respondents of lower educational background are 'Willing' instead of 'Very much willing' to donate some money regularly for any environmental movement: general or local. Their willingness stands against the conventional wisdom of elitist bias associated with modern environmentalism.

Table 8.6: Zonal Variation in Willingness to Contribute Money for Environmental Movement: General *versus* Local by Educational Levels

Education	*Willingness to Contribute Money*								*Total*
	Unwilling		*Undecided*		*Willing*		*Very much willing*		
	General	*Local*	*General*	*Local*	*General*	*Local*	*General*	*Local*	*N*
High Pollution Zone									
Lower (Up to Secondary)	21.3	**7.41**	24.07	**19.44**	52.78	**63.89**	1.85	**9.26**	108
Higher (Graduate and above)	10.2	**4.08**	14.29	**12.24**	53.06	**48.98**	22.45	**34.69**	49
Total	17.8	**6.37**	21.02	**17.2**	52.87	**59.24**	8.28	**17.2**	157
Low Pollution Zone									
Lower (Up to Secondary)	26.4	**10.38**	15.09	**8.49**	56.6	**74.53**	1.89	**6.60**	106
Higher (Graduate and above)	9.52	**3.18**	22.22	**20.63**	46.03	**44.44**	22.22	**31.75**	63
Total	20.1	**7.69**	17.75	**13.02**	52.66	**63.31**	9.47	**15.98**	169

Table 8.7: Zonal Variation in Willingness to Participate in Environmental Movement: General *versus* Local by Income

Income	*Willingness to Participate*								*Total*
	Unwilling		*Undecided*		*Willing*		*Very much willing*		
	General	*Local*	*General*	*Local*	*General*	*Local*	*General*	*Local*	*N*
High Pollution Zone									
Lower (Up to Rs. 2000)	15.3	**17.65**	20	**10.59**	24.71	**11.76**	40	**60**	85
Higher (Rs. 2001 and above)	6.73	**3.85**	14.42	**9.62**	32.69	**16.35**	46.15	**70.19**	104
Total	10.6	**10.05**	16.93	**10.05**	29.1	**14.29**	43.39	**65.61**	189
Low Pollution Zone									
Lower (Up to Rs. 2000)	32.9	**12.94**	10.59	**14.12**	35.29	**30.59**	21.18	**42.35**	85
Higher (Rs. 2001 and above)	12.9	**12.87**	18.81	**9.90**	37.62	**21.78**	30.69	**55.45**	101
Total	22	**12.9**	15.05	**11.83**	36.56	**25.81**	26.34	**49.46**	186

Table 8.7 perhaps most strongly indicates the support for local environmental movement as the modal response of support for local movement falls into the 'Very much willing' category against the modal support for general movement plunges into the 'Willing' group. However, among the respondents very much willing to support environmental movements, income differential of support base is also noteworthy as the upper income category respondents pledge to support such movement in greater proportion irrespective of their zone of residence, than their lower income group counterparts. Once again, this supports the conventional wisdom of an elitist bias of environmentalism. Nevertheless, the importance of zonal difference also remains clear as the proportion of respondents very much willing to support local as well as general environmental movement is noticeably higher in High Pollution zone compared to Low, supporting the stake of objective existence of pollution as an important determinant of anti-pollution movement support. For more specific indices of support, let us take a glance at Tables 8.8 and 8.9.

Table 8.8: Zonal Variation in Willingness to Devote Time for Environmental Movement: General *versus* Local by Income

Income	*Willingness to Devote Time*								*Total*
	Unwilling		*Undecided*		*Willing*		*Very much willing*		
	General	*Local*	*General*	*Local*	*General*	*Local*	*General*	*Local*	*N*
High Pollution Zone									
Lower (Up to Rs. 2000)	31.8	**12.94**	30.59	**25.88**	34.12	**55.29**	3.53	**5.88**	85
Higher (Rs. 2001 and above)	14.4	**9.62**	34.62	**28.85**	44.23	**56.73**	6.73	**4.81**	104
Total	22.2	**11.11**	32.8	**27.51**	39.68	**56.08**	5.29	**5.29**	189
Low Pollution Zone									
Lower (Up to Rs. 2000)	45.9	**24.71**	18.82	**15.29**	34.12	**56.47**	1.18	**3.53**	85
Higher (Rs. 2001 and above)	19.8	**9.90**	36.63	**27.72**	35.64	**54.46**	7.92	**7.92**	101
Total	31.7	**16.67**	28.49	**22.04**	34.95	**55.38**	4.84	**5.91**	186

Willingness to devote time for environmental mobilization is cross-tabulated against income, both for High and Low Pollution zones in Table 8.8. While modal category of response is 'Willing' as earlier, interestingly, the respondents of lower income category either parallels or supersedes their higher income counterparts in both the zones, particularly regarding participation in local movements. This goes against the conventional wisdom that environmentalism is a middle/upper-middle class phenomenon, so long as giving time to movement is concerned. Whatsoever, the zone difference for the respondents of 'Very much willing' category indicates that the higher income group in both the zones are equivocal in their support to general as well as local environmental movements. Thus, in line with the conventional wisdom, these middle/upper middle class elites, perhaps, exhibit greater environmental maturity and they perhaps reflect higher environmental concern as they view the local and general environmental mobilizations as equals and very much willing to support those.

Table 8.9: Zonal Variation in Willingness to Contribute Money for Environmental Movement: General *versus* Local by Income

Income	*Willingness to Contribute Money*								*Total*
	Unwilling		*Undecided*		*Willing*		*Very much willing*		
	General	*Local*	*General*	*Local*	*General*	*Local*	*General*	*Local*	*N*
High Pollution Zone									
Lower (Up to Rs. 2000)	24.7	**7.06**	25.88	**22.35**	49.41	**65.88**	0	**4.71**	85
Higher (Rs. 2001 and above)	8.65	**5.77**	20.19	**15.38**	55.77	**50.96**	15.38	**27.88**	104
Total	15.9	**6.35**	22.75	**18.52**	52.91	**57.67**	8.47	**17.46**	189
Low Pollution Zone									
Lower (Up to Rs. 2000)	23.5	**9.41**	18.82	**9.41**	56.47	**78.82**	1.18	**2.35**	85
Higher (Rs. 2001 and above)	19.8	**8.91**	15.84	**14.85**	46.53	**46.53**	17.82	**29.7**	101
Total	21.5	**9.14**	17.2	**12.37**	51.08	**61.29**	10.22	**17.2**	186

Table 8.9 also reports 'Willing' as the modal category of response and a bias in favour of local in both the zones. More importantly, the

respondents of lower income category exhibit a deep disinterest in general environmental movement, signifying their inclination to view environmental crisis as a local problem only. This might be related to general lack of environmental awareness often associated with their disadvantaged socio-economic standing.

While this is in tune with the conventional wisdom, their larger presence in the 'Willing' category indicating their marginal support both for general and local movements, particularly in the Low Pollution zone, stands against the conventional wisdom. Thus, over all income emerges as an important differential besides the objective existence of pollution in one's local environment as an important determinant of support to environmental movement.

The CPNC Movement

Present study reveals that the respondents of high Socio-Economic Status (SES) group (with relatively higher income, property and higher education, general awareness, social participation etc.) are more aware regarding pollution and its harmful effects, than the respondents of lower socio-economic-status group. However, politics possibly plays an important role in this context as the ruling elites argue that the problem of pollution is not of very serious dimension, whereas the activists of the opposition camp tries to magnify the actual problem in heir locality.

People of both the high and low SES have favourable attitude towards the environmental movement of CPNC. However, the supporters of the party in state power, irrespective of their SES, are against the popular belief that pollution is creating a lot of problems for the local community. They are also against highlighting the role of CPNC, which is, as per its declared objectives, politically neutral but which, in its protest movements against the Government's negligence in taking any serious measure to check pollution, often assumes the character of an anti-ruling party organization.

Generally, the participants and organizers of the CPNC are people of higher SES group. To them the environmental pollution is the most important problem in the area. However, the people of lower SES are more concerned regarding their poverty and livelihood than the environmental pollution. To them, environmental pollution is only one problem amongst others. They lay certain stress on adaptation up to an extent with the problem, and at the same time they also extend support to the protest movements.

Now the question is can the social movement led by the CPNC be explained in terms of value change and post-materialist thesis? Can it be called a new social movement? We shall deal with each question separately.

The post-materialist argument as put forth by Inglehart (1971 and 1977) is that with rising standards of living in the developed world many peoples' material wants are being satisfied. Consequently people are becoming more and more oriented towards less acquisitive and non-material goals to do with factors such as quality of life, intellectual and spiritual development, political rights, participation and the environment. On the other hand, Martell (1994) has argued that the post-materialism could be the result rather than the cause of the rise of new social movements giving high priority to post-material issues. Again, rise of environmentalism along with other new social movement issues may be due to an increased profile for environmental issues in the media or education. Both the original post-materialist thesis and these other possible explanations suggest, meanwhile that it is appropriate to see the basis of values in social and environmental processes rather than to see concern or action as deriving simply from values.

India, being predominantly a developing country with low standard of living in general and with a lot of material wants of her people, has not yet entered generally in the post-material stage of development. Value change and post-materialist thesis seemingly fall short of explaining the movement of CPNC. In fact the survey reveals that the leadership of the movement consists primarily of people of different standards of living, of different class positions: the educated unemployed persons, doctors, teachers, retired persons and a few workers of the factories of the locality. Again, many of them are politically excluded elites who strive for leadership outside the established political circles, and a sizeable faction of them is of young age. In short, the CPNC activists consist of two groups of people: individuals concentrated in middle class occupations in the non-market sector, especially in service professions in education (such as teachers), social work and health (doctors), in the one hand and educated unemployed youths and retired persons, on the other.

Consequently, three specific factors, possibly, are responsible for the emergence of the CPNC movement. First, concern for a cleaner physical environment and better quality of life of the first group (consisting of doctors, teachers, retired persons etc.), as well as their political exclusion from and striving for leadership outside of the established political circles. Second, the frustration and grievance of the educated unemployed people have most possibly encouraged them to be got involved in such a movement that is anti-state, anti-establishment in nature. Their frustration and grievance, in other words, most possibly, are routed into anti-state environmental movement. Finally, there still remains an obvious explanation yet the most poorly considered in the environmental sociology that environmentalism is a response to environmental problems (Martell, 1994). This is very much applicable to the CPNC movement. The movement is concentrated, as is said earlier, in the Chandrahati village where almost all the CPNC activists reside. And this area is the most severely affected by environmental pollution in terms of magnitude and duration. On the other hand the movement is almost absent in the Majherchar village where the intensity of the problem is much less. All these, at least partially, confirm the proposition that environmentalism is a response to the environmental problems.

Let us come to the second question, i.e., can the movement of the CPNC be termed as a 'New Social Movement'? The chief features of the NSMs in contrast to old social movements, according to Martell (1994) are as following:

(1) *Location*: Old social movements tend to be located in the polity in political parties, whereas the new social movements are autonomous movements outside conventional political institutions.

(2) *Aims*: The aims of old social movements are to secure political representation, legislative political reform etc.; the new social movements, however, want to defend civil society against political power and redefine culture and life-style in civil society rather than pursuing legislative change through the State.

(3) *Organization*: Old social movements are formal and hierarchical whereas the NSMs go for informal or

unstructured organization or 'networks' built up from grass-roots participation rather than structures of authority.

(4) *Medium of change*: Old social movements are oriented towards participation in political institutions, whereas the NSMs go for newer and more innovative forms of direct action and cultural politics.

Another can be added with the above-mentioned four features, and this is regarding the nature of the participants.

(5) *Participants*: Nationalists, peasants, workers etc. are the participants of the old social movements whereas in NSMs, participants are mainly of different social categories; namely, unemployed educated youths, retired persons, persons of white-collar middle class occupations of the non-market sectors (such as doctors, teachers, voluntary social workers etc.), educated middle class housewives etc.

Moreover, new social movements often are localized and temporary in nature.

In certain respects, considering the above-mentioned features, the movement of the CPNC comes very close to the ideal type of a new social movement indeed; however, taken as a whole possibly it cannot be termed as a new social movement.

First, in terms of location, the movement of the CPNC is a new social movement undeniably as it is an autonomous movement outside the conventional political institutions and parties. Perhaps this is the single most important factor for the weakness of the movement at the same time as, allegedly, at every step it has to face hard resistance from the established political parties.

Second, in terms of aims, though the CPNC movement tries to defend civil society against political power and in that count is a new social movement, still as from time to time it has taken recourse to legislative changes through judiciary and the state, it can not be termed as a new social movement.

Third, from the organizational viewpoint, the CPNC movement is a new social movement indeed. This is evident from the informal, unstructured, non-hierarchical nature of its organization. Though there is a core group, which takes initiative for any organizational, or movement programme, it always takes decision democratically and stresses on grassroots participation. However, the study reveals that it suffers from alienation to some extent from the illiterate people of lower socio-economic strata of the area and in this sense it deviates from the paradigm of new social movement. Such an organizational structure, it seems, is another factor for the weakness of the movement.

Fourth, the CPNC movement, in terms of the 'medium of change' has neither gone for newer and innovative forms of direct action nor pursued the path of cultural politics. CPNC activists published several leaflets and pamphlets, organized rallies, meetings, *dharnas* (sit-in demonstrations), conducted mass signature collection drives, pursued legal battles etc., in the same line with different political movements of the established political parties or old social movements. Thus the medium of change of the CPNC movement does not correspond to the new social movement scheme.

Fifth, in terms of participants, the CPNC movement does not strictly adhere to the new social movement characteristics. The participants of the CPNC movement are mainly educated unemployed youths, persons of white-collar occupations of the non-market sectors (such as doctors, teachers and voluntary social workers), retired persons etc. and in this sense it fits to the new social movement paradigm. However, the workers of the factories of this area—the people of lower middle class occupations of the market sectors—also are its participants and here is a little bit deviation from the new social movement paradigm.

Finally, the temporary and extremely localized nature of the CPNC movement resembles the feature of a typical new social movement. However, this may be a possible cause of weakness of the movement too. Hence, an overall assessment of the CPNC movement exhibits that the movement is not a new social movement in strict sense. In fact, as mentioned earlier, environmental movements throughout modern India are generally anti-state in nature and the major participants of these movements are the lower class rural people.

Beyond doubt, environmental concern in the developing countries, in the context of their highest priority for industrialization and other developmental activities, is to some extent luxurious. Poverty is supposedly the most important pollutant in these countries and development happens to be the distinct answer. However, in course of development, environment has to be safeguarded as far as possible. Our scientists, natural and social together, have to find out an optimum balance in development-environment debate. Sustainable development is required to build up a sustainable society. Spread of education and concern for a better quality of life and a greater concern of environmental issues is, probably, the basic solution. People with greater environmental awareness will eventually create a situation where environmental consequences of developmental activities could hardly be ignored.

REFERENCES

Bandyopadhyay, J. "Chipko Movement: Of Floated Myths and Floated Realities", *Economic and Political Weekly*, April 10, 34: 1999. pp. 880-82.

Bapat, J. "Determinants of Ecological Consciousness: A Path Analysis", *Journal of Indian Anthropological Society* 23, 1988. pp. 113-129.

Baviskar, A. *In the Belly of the River: Tribal Conflicts over Development in the Narmada Valley.* Oxford University Press, New Delhi, 1995.

______ (a) " Who Speaks for the Victims?", *Seminar,* 451, 1997. pp. 59-61.

______ (b) "Ecology and Development in India: A Field and its Future", *Sociological Bulletin,* 46, 1997. pp. 193-207.

Beck, U. *Risk Society: Toward a New Modernity,* Sage, London, 1992.

Beteille, A. "Hierarchical and Competitive Inequality", *Sociological Bulletin,* 51(1), 2002. pp. 3-27.

Brechin, S.R. and W. Kempton "Global Environmentalism: A Challenge to the Post materialism thesis?", *Social Science Quarterly,* 75, 1994. pp. 245-269.

Buttel, F.H. "Age and Environmental Concern: A Multivariate Analysis", *Youth and Society*, 10, 1979. pp. 237-256.

Buttel, F.H. "New Directions in Environmental Sociology", *Annual Review of Sociology,* 13, 1987. pp. 465-488.

______ "Environmentalization: Origins, Processes, and Implications for Rural Social Change", *Rural Sociology,* 57, 1992. pp. 1-27.

Buttel, F.H. and W.L. Flinn "The Structure of Support for the Environmental Movement: 1968-1970", *Rural Sociology,* 39, 1974. pp. 56-69.

Buttel, F.H. and W.L. Flinn "Environmental Politics: The Structuring of Partisan and Ideological Cleavages in Mass Environmental Attitudes", *The Sociological Quarterly,*17, 1976. pp. 477-490.

Chapman, G., K. Kumar, C. Fraser and I. Gaber, *Environmentalism and the Mass Media: the North-South Divide,* Routledge, London, 1997.

Chatterjee, D. P 'Environmentalism and Appropriate Technology", *Socialist Perspective* 26, 1999. pp. 163-178.

_________ (2004) "Environmental Movement in a Local Context: The Case of Anti-Pollution Movement in a Small Locality of West Bengal." *Guru Nanak Journal of Sociology,* 25, 2004. pp. 35-56.

_________ "Oriental Disadvantage *versus* Occidental Exuberance: Appraising Environmental Concern in India—A Case Study in a Local Context", *International Sociology,* 23, 2008. pp. 05-35.

CSE "Small and Medium Towns" in Anil Agarwal, Sunita Narain and Srabani Sen (eds.) *State of India's Environment: The Citizens' Fifth Report (Part I: National Overview)* 1995, pp. 209. New Delhi: Centre for Science and Environment.

Dasgupta, B. "The Environment Debate: Some Issues and Trends", *Economic and Political Weekly,* Annual Number, February, 1978: 385-400.

Dunlap, R.E. and A.G Mertig "Global Concern for the Environment: Is Affluence a Prerequisite?", *Journal of Social Issues,* 51, 1995. pp. 122-37.

Dunlap, R.E., G.H. Gallup, and A. M. Gallup (1993). "Of Global Concern: Results of the Health of the Planet Survey", *Environment,* 35, 1993. pp. 7-15 and 33-39.

Desai, A. *Environmental Perception: The Human Factor in Urban Planning,* Ashish Publishing House, New Delhi, 1985.

D'Souza, V.S. "Beyond Fifty Years of India's Independence: Challenges to Scientific Sociology", *Sociological Bulletin,* 47, 1998. pp. 9-15.

Franzen, A. "Environmental Attitudes in International Comparison: An Analysis of the ISSP Surveys 1993 and 2000", *Social Science Quarterly,* 84, 2003. pp. 297-308

Foster, J.B. "Environmental Sociology and the Environmental Revolution: A 25th Anniversary Assessment", *Organization and Environment,* 15,2002. pp. 55-58.

Gadgil, M. and R. Guha "Ecological Conflicts and the Environmental Movement in India" in Dharam Ghai (ed.) *Development and Environment: Sustaining People and Nature,* Blackwell Publishers, Oxford, 1994.

Gale, R.P. "The Environmental Movement and the Left: Antagonists or Allies?", *Sociological Inquiry,* 53, 1983, pp. 179-199.

Giddens, A. *The Consequences of Modernity,* Polity Press, Cambridge, 1990.

Giddens, A. *Modernity and Self-Identity,* Polity Press, Cambridge, 1991.

Gillham, P.F. "Participation in the Environmental Movement Analysis of the European Union", *International Sociology,* 23, 2008. pp. 67-93.

(GOI, N.D.) Year-End-Review, Home. Available online at: http://pib.myiris.com/review2000/article.php3?fl=home

Guha, R. *The Unquiet Woods: Ecological Change and Peasant Resistance in the Himalaya,* Oxford University Press, New Delhi, 1989.

_________ "Prehistory of Indian Environmentalism", *Economic and Political Weekly* January 4-11, 1992: 57-64.

Habermas, J. "New Social Movements", *TELOS,* 49, 1981. pp. 33-37

Hansen, A. (1993) 'Introduction' in Anders Hansen (ed.) *The Mass Media and Environmental Issues,* Leicester University Press Leicester, London and New York, 1993.

Inglehart, R. "The Silent Revolution in Europe: Intergenerational Change in Post-industrial Societies", *American Political Science Review,* 65, 1971. pp. 990-1017.

________ *The Silent Revolution: Changing Values and Political Styles Among Western Publics*, Princeton University Press, Princeton, 1977.

________ "The Persistence of Materialist and Post-materialist Value Orientations: Comments on Van Deth's Analysis", *European Journal of Political Research,* 11, 1983. pp. 81-91.

Jackson, C. "Environmentalism and Gender Interests in the Third World", *Development and Change,* 24, 1993. pp. 649-77.

Jathol, P.S. "Environmental Pollution and Village Life", *Social Change*, 21, 1991. pp. 40-56.

Jayal, N.G. "Uttaranchal: Same Wine, Same Bottle, New Label?", *Economic and Political Weekly*, December 2-8, 35, 2000. pp. 4311-14.

Jehlicka, P. "Environmentalism in Europe", Paper to British Sociological Association Conference, April 1992.

Jehlicka, P. and T. Kostelecky "The Greens in Czechoslovakia". Paper to Joint Session of Workshops, University of Essex. March 1991.

Karan, P.P. "Public Awareness of Environmental Problems in Calcutta Metropolitan Area", *The National Geographic Journal of India*, 26,1980. pp. 29-34.

Karan, P.P. "Environmental Movements in India", *Geographical Review*, 84, 1994. pp. 32-41.

Kumra, V.K. *Kanpur City: A Study in Environmental Pollution*, Tara Book Agency, Varanasi, 1982.

Lowe, P.D. and W. Rudig "Political Ecology and the Social Sciences—The State of the Art", *British Journal of Political Science*, 16, 1986. pp. 513-50.

Lundy, P. "Fragmented Community Action or New Social Movement: A Study of Environmentalism in Jamaica", *International Sociology*, 14, 1999. pp. 83-102.

Martell, L. *Ecology and Society: An Introduction*, Polity Press, Cambridge, 1994.

Mawdsley, E. "After Chipko: From Environment to Region in Uttaranchal", *The Journal of Peasant Studies*, 25, 1998. pp. 36-54.

Michelson, W.H. *Man and His Urban Environment*. Reading, Mass.: Addison-Wesley, 1976.

Munshi, I. "Environment in Sociological Theory", *Sociological Bulletin*, 49, 2000. pp. 253-66.

Oommen, T.K. "Technology and Environment: Towards a Viable Policy Perspective" in *State and Society in India*, Sage Publications, New Delhi, 1990. pp. 139-153.

Pichardo, N. A. "New Social Movements: A Critical Review", *Annual Review of Sociology*, 23, 1997. pp. 411-30.

Rannikko, P. "Local Environmental Conflicts and the Change in Environmental Consciousness", *Acta Sociologica*, 39, 1996. pp. 57-72

Rootes, C. (a) "Environmental Movements and Green Parties in Western and Eastern Europe" in M. Redclift, and G. Woodgate edited *The International Handbook of Environmental Sociology*, Edward Elgar, Cheltenham, 1997.

_________ (b) "Social Movements and Politics", *African Studies*, 56, 1997. pp. 67-95.

_________ (a) "Acting Locally: The Character, Contexts and Significance of Local Environmental Mobilisations", *Environmental Politics*, 16, 2007. pp. 722-41.

_________ (b) "Global Civil Society: Just A Noble Aspiration, Or An Emergent Reality?" in *Mapping the public policy landscape: From local to global, ESRC Seminar Series*, 2007. ESRC, pp.14-21. Swindon. (Available online at: www.esrc.ac.uk/ESRCInf°Centre/Imagesfrom_local_to_global_tcm6-21279.pdf)

Sethi, H. "Survival and Democracy: Ecological Struggles in India" in Ponna Wignaraja (ed.) *New Social Movements in the South: Empowering the People*, Vistaar Publications, New Delhi, 1993.

Shiva, V. "Ecology Movements in India", *Alternatives*, 11, 1986. pp. 255-273.

_________ *Staying Alive: Women, Ecology and Survival in India*, Kali for Women, New Delhi, 1988.

_________ *Ecology and the Politics of Survival: Conflicts over Natural Resources in India*, Sage Publications, New Delhi, 1991.

_________ "North-south Conflicts", *Frontier*, 24, 1992. pp. 10-14.

_________ "Conflicts of Global Ecology: Environmental Activism in a Period of Global Reach", *Alternatives*, 19, 1994. pp. 195-207.

Singh, R. *Social Movements, Old and New: A Post-modernist Critique*, Thousand Oaks, Sage Publications, London, 2001.

Srinivas, M.N. *Caste in Modern India and Other Essays*, Asia Publishing House, Bombay, 1962.

van der Heijden, H. A. "Globalization, Environmental Movements, and International Political Opportunity Structures", *Organization and Environment*, 19, 2006. pp. 28-45.

Van Liere, K.D. and R.E. Dunlap "The Social Bases of Environmental Concern: A Review of Hypotheses, Explanations, and Empirical Evidence", *Public Opinion Quarterly*, 44, 1980. pp. 181-97.

Venugopal, P. Plachimada: Water vs. Coca Cola, *The Hindu Survey of the Environment, 2003*: 87-89

Vorkinn, M. and H. Riese 'Environmental Concern in a Local Context: The Significance of Place Attachment', *Environment and Behaviour*, 33, 2001. pp. 249-63.

Yanitsky, O. "The Environmental Movement in a Hostile Context: The Case of Russia", *International Sociology*, 14, 1999, pp. 157-172.

Yearley, S. *The Green Case: A Sociology of Environmental Issues, Arguments and Politics*. London: Harper-Collins, London, 1991.

_________ "Environmental Social Science and the Distinction Between Resource use and Industrial Pollution: Reflections on an International Comparative Study", *Ambiente and Sociedade* 8, 2005. pp. 11-26.

9

Urban Environment and its Governance

JASLEEN KEWLANI and MANMOHAN SINGH GILL

Social change is inevitable in nature and it reinforces multiple forms of alterations thereon. The transitional situation leads to emergence of many new issues. These issues emerge as great challenges for the governance and administration. Some prominent socio-demographic alterations occurred in Indian scene also, which were the result of urbanization. Urbanization is such a marked trend, which has continued evolving, since the later half of twentieth century. A mass exodus of rural population can be witnessed in census and even in common civic visions of Indian cities. Cities are labeled as 'cardinal regions' for economic growth and attract the rural population with the 'pull power' of their development paradigm, but an unapologetic and unhesitant influx of rural people to urban areas along with their countryside survival pattern, causes many problems for governance in cities. It implies that in this process of geographical shift; thought process, living habits and abstract versions of rural folk remain intact, but this folk expects that all the amenities so far being relished by the urban dwellers should be equally shared with the new entrants. On the other hand, the absorbing capacity of the city and its infrastructure do not enhance at par with the scale of urbanization. This urban paradox puts stress on the need to improve urban administration, in order to cope with some cogent issues like housing, pollution control, environmental maintenance, social adaptability, health enhancement, etc. These targets can be achieved by adopting a "Sustainable Human Development Pattern" which involves socio-economic banking of individuals and communities already residing in urban areas and the recently arrived alien communities also. This notion puts forward an assertive view that urban environment can be managed and improved with efficacious regulation of services being provided in the cities.

The paper aims at discussing cogent issues related to urban environment management and services being provided in cities. At the same time, it endevours to highlight those aspects, which led to further deterioration of urban environments as a result of environment and management of services. Another target of the paper is to present the rhetoric of community participation, through NGOs and Community Based Organizations (CBOs), which provides a facilitated platform for management of environment in cities. In all, there are two main sections of the paper. (1) Environmental Management and Services provided in the cities; and (2) The role of Urban Local Bodies (ULBs) and importance of community participation with a special reference to NGOs and CBOs functioning in urban environment management domain.

Secondary sources have been used to accomplish the needed information. A large quantum of data has been availed from Census of India, 2001. Other than this, many United Nations' programmes have been explained, which are specifically related with urban environment and its management. Some vital national as well as international demographic reports have been consulted.

SECTION I

Environment Management and Services provided in the Cities

The management of Urban Environment requires a well planned strategy. Environment is endowed with a "carrying capacity" which implies a limitless power of the natural environment to act as a reservoir of resources which supports human's unending consumption and also as a sink which assimilates the waste products and their adverse environmental effects. United Nations Conference on Environment and Development (UNCED), has adopted Sustainable Development Settlement, as its main agenda of action, but not only India, rather a number of Third World countries are yet to operationalize this agenda in their own planning strategy which is not possible until and unless the potential of 'carrying capacity' of environment is realized and the environmental decay is reversed. If it is done, it will surely lay provisions for improvement in basic services and fundamental infrastructure. In absence of these credentials, urban population of India or of any Third World developing countries is leading a marginalized life and is unable to realize its own potential or to contribute the worth to the communities around it. In India three main cities like Mumbai, Kolkata and Delhi, between 40-50 per cent of the dwellers live in informal, congested, often illegal and unhealthy slums, which are not served by city facilities. Not only slums but other areas of the cities also face an underprivileged pattern of survival due to lack of efficacious environment management. No doubt, providing all facilities to everyone is a colossal task but at least a minimal level improvement needs to be initiated. Environmental impacts do not leave any domain of life untouched as an effect of environmental changes. Adverse or positive impacts are must to occur, as an effect of environmental changes, especially in case of India where large scale urbanization is taking place. Environmental management becomes vital, as it Affects well-being of the residents to a large extent. The situation becomes even more alarming, when urbanization occurs in such an uncontrolled manner as shown in Table 9.1, which shows the trend of urbanization in India.

Urban Environment and Over-crowding

Population statistics sometimes get dominated by the density factor. Excess density of population reinforces the alarming ramification. Over-crowding leads to emergence of many socio-medical issues. 'Mumbai' the financial capital of India, is a good example in this context. This city covers approximately 457.71 sq.km. of land. Immigration scene of this city presents a bombardic influx of people per day. 'Dharavi' the biggest Asia's slum is also a part of this geological unit. Daily garbage generation ranges more than 3.20 tons, leading to 9 to 20 per cent lungs disease incidence. Noise pollution is 72 decibels in many areas. According to Jayapalan (2002) pollution of air in slum areas is equal to smoking of 30 cigarettes per day. These are the chemical effects of over-crowding in 'Mumbai'. Moving slight away from the main theme of the paper a fact well related to that of over-crowding is being presented and has moral ramifications associated with any policy of sustainable development. A study

Table 9.1: Trend and Pattern of Urbanization across States in India

States/Union Territory	*Urban Population 2001(per cent)*	*Growth Rate*	
		1981-91	*1991-2001*
India	27.8	3.1	2.7
Andhra Pradesh	27.1	3.6	1.4
Arunachal Pradesh	20.4	9.9	7.0
Assam	12.7	3.3	3.1
Bihar (undivided)	13.4	2.7	2.6
Chandigarh	89.8	3.1	3.4
Delhi	93.0	3.8	4.1
Goa	49.8	3.9	3.3
Gujarat	37.3	2.9	2.8
Haryana	29.0	3.6	4.1
Himachal Pradesh	9.8	3.2	2.8
Jammu and Kashmir	24.9	3.8	3.1
Karnataka	34.0	2.5	2.5
Kerala	26.0	4.8	0.7
Madhya Pradesh (undivided)	25.0	3.7	2.8
Maharashtra	42.4	3.3	2.9
Manipur	23.9	3.0	1.2
Meghalaya	19.6	3.1	3.1
Mizoram	49.5	9.5	3.3
Nagaland	17.7	5.5	3.3
Orissa	15.0	3.1	2.6
Punjab	34.0	2.5	3.2
Rajasthan	23.4	3.3	2.7
Sikkim	11.1	-3.2	4.8
Tamil Nadu	43.9	1.8	3.6
Tripura	17.0	6.2	2.5
Uttar Pradesh (undivided)	20.8	3.3	2.8
West Bengal	28.0	2.6	1.8

Sources : Census of India; 2001.

(Panda, 2002) was conducted in 'Brihan', i.e. Greater Mumbai which has about 350 households. According to the residents of this area, AIDS infections are the result of over-crowding. In this region more than 16-18 members of family live in one large room due to excessively high density of population and lesser housing settlements. It retards the privacy factor. Couples are not able to fulfil their sexual needs, for which males move out to seek sex services of Commercial Sex Workers (CSWs) which lead to an increase in HIV/AIDS infections. But gendered discrimination instincts cannot be ruled out in males because they blame the Commercial Sex Workers (CSWs) for providing unsafe sex practices leading to HIV/AIDS infections. It can be well analyzed with various examples that how over-crowding in the city leads to social issues, medical problem and even in chemical deterioration of atmosphere. A thick density of population in an area leaves alarming repercussions on multiple aspects of survival. In Table 9.2, the data concerning density of some important cities of India is being given.

Table 9.2: Density of Some Important Cities of India

Sl. No.	*Name of Cities*	*Population*	*Density*
1.	Amritsar	3,374,207	603
2.	Jalandhar	1,953,508	742
3.	Ludhiana	3,030,352	804
4.	Chandigarh	900,914	7,903
5.	Ambala	1,013,660	644
6.	Delhi	13,782,976	9,294
7.	New Delhi	171,806	4,909
8.	Ghaziabad	3,289,540	1,682
9.	Kanpur	4,137,489	1,366
10.	Kolkata	4,580,544	24,760
11.	Mumbai (Sub-Urban)	8,587,561	19,255
12.	Mumbai	3,326,837	21,190
13.	Hyderabad	3,686,460	16,988
14.	Chennai	4,216,268	24,231

Source : Censsu of India, 2001.

A high density of population leaves adverse effects on the issues like solid waste management, drinking water, health sector and other related environmental aspects. In all, the whole infrastructure gets threatened if density of population in an area exceeds the reasonable limits.

(i) Urban Environment and Infrastructure

The Indian Infrastructure Report, commonly known as 'Rakesh Mohan Committee Report' estimated that the infrastructure investment would increase from about Rs.4,000-4,500 billion

(US$115 to 130 billion) during 1995-2000, to about Rs. 7,500 billion (US$215 billion) during 2001-06. This committee was constituted under the Ministry of Finance, Government of India. Its assertion shows that the need to develop the infrastructure has increased greatly. The 'Global Scenario Group', a popular world level statistical organization explored the expected future urbanization trends up to the year 2050. According to its results, as compared to 1995, the total population in developing regions will almost be double to about 8 billion. At present also, most urban dwellers live in developing countries. The population in the developed part of the world will receive a slight increase, which will be up to less than 1.5 billion and about 90 per cent of the population will live in urban areas. Hence, the report concludes that the rate of urbanization in less advanced nations will almost double in this 50-years period to about 70 per cent (World Development Report, 2003). This anticipated rate can be implied in Indian scene also which certainly poses innumerable challenges in context of infrastructure, its maintenance and development. Moreover, it may lead to distorting impacts on environmental perspectives. Development of metropolitan areas at sub-regional or even at regional level plays a dominant role in future land-use patterns. On the other hand, investment on roads, railways, water ways, harbours and airport (important aspects of Infrastructure Development), certainly replace some or other type environmental unit like forests, parks, water bodies, lagoons etc. Lack of will to implement strict land use plans, instead of improving and developing the survival internalities, destroys the already available given cover entities in the name of development. And all this deteriorates the environment of the urban areas. This view was supported by UNEP's "Global Environment Outlook, 2000". Its finding made the need for local action very ocular. Further, if environment is not well protected, the emission of sulphur and nitrogen compound to the air, will at least triple in the next three decades in continental Asia. To stabilize these levels of emission, Comprehensive Urban Policy is needed. This Comprehensive Urban Policy incorporates introduction of clean technologies in energy production and transport, fuel switching, increasing transportation efficiency, and the promotion of investment in public transport (UNEP, 2000).

According to an assessment made by the World Bank, India's annual environmental damages conservatively amounted to $9.7 billion or 4.5 per cent of the Gross Domestic Product (GDP) in 1992, as compared to 2.6 per cent of China in the same year. Six cities in India, namely, Mumbai, Kolkata, Delhi (among 10 most populated cities in the world), Ahmedabad, Kanpur and Nagpur, have annual average total suspended particulates, which are nearly three times of the standard laid by the World Health Organization (WHO), leading to excess respiratory problems in India. If it gets checked, according to WHO, it will be helpful in saving at least 37,000 lives annually. Unprecedented and incessant urbanization has aggravated the environmental crisis. Burning of biomass in cities, burning of wood and coal for domestic cooking and use of unclean technologies are some common urban trends in India, which are leading to mounting of air pollution in India and contribute 8 per cent to the global warming. India's Urban Environment needs to be improved. In the following part of the paper, some important services in context of environment management will be assessed.

(ii) Water : Its Quality and Impact on Health

Inadequate water disposal has unavoidable negative impacts, which further get aggravated due to weak organizational capacity for collective action. As compared to conventional methods, the condominal sewerage reduces investment costs by half, but requires a strong commitment of household to maintain it collectively (World Development Report, 2003). Issue of safe drinking water, least wastage of pure water etc., are important issues in maintenance of urban environment. A broader look can be had over the issues with the help of Table 9.3, which presents environment, health, welfare and living conditions: of various cities of the world, divided on the basis of various city product categories. Four product categories have been enlisted, i.e. low product, lower-middle, upper-middle and high city product. The calculation of city product implies a city's average income calculated by natural accounts method. Low income defined as city product below $750 per capita a year; lower-middle as $-751-2,499, upper-middle as $2,500-9,999; high as above $10,000.

Table 9.3: Environmental Health, Welfare, and Living Condition vary by City Product

Indicators	*City Product Category*			
	Low	*Lower Middle*	*Upper Middle*	*High*
Household connections (per cent) large				
Water	48.0	78.8	92.9	99.9
Sewerage	45.6	68.7	84.3	99.9
Electricity	72.3	93.6	95.0	100.0
Water consumption large				
Litres per person per day in All settlements	88	161	232	247
Waste water treated				
Per cent treated	29.4	56.7	68.2	97.4
Solid waste disposal (per cent) large				
Sanitary landfill or incinerated	30.7	41.4	37.6	77.7
Other (Open Dump, Recycled, Burned)	65.9	58.3	62.2	22.3
Under-five Mortality (Per-1,000)	104.2	39.7	25.8	5.2
Household below the locally/ defined poverty line (per cent)	31.7	23.2	16.0	6.9
Sample Size (cities)	49	36	25	20

Source: United Nation + HABITAT, Global Urban Indicators Database (1998).

The table clearly elaborates the scene of various health related environmental indicators such as water connection, consumption, and waste water management. Various statistical organizations have proved that water supply networks result in high losses of water, incurred due to inadequate transmission and distribution pattern. The loss ranges from 25 per cent to 50 per cent. Back siphoning of water due to low pressures leads to the contamination of water, which further creates a situation where untreated sewerage becomes primary source of water pollution (Ghosh, 2003). On 31st March,1997, 49 per cent of the urban population had sewage and sanitation facilities and it improved to 55 per cent in the next three years, i.e. up to 2000. Out of 300 class-I towns with population of 100,000 or more, about 70 per cent have partial sewerage and sewage treatment facilities (HUDCO and Habitat, 2001). In all the cases, the stress is laid on improving the facilities but not on maintaining these in terms of quality and distribution. In Calcutta during monsoon, sewerage water enters the portable water system leading to a number of health hazards and no serious attempt is made to improve upon this system. This explains lack of managerial attitude among the governing bodies of the cities.

Moreover, the urbanization process associated with growth of industries and trade too contributes towards environmental pollution/ toxication of groundwater and soil. Industrial chemicals released by iron and steel produces contribute to nearly 70 per cent of the effluent released, on the contrary 20 per cent of these chemicals are used for the industrial ouput. It presents a great paradox. This unregulated disposal of pollution occurs through drains joining into rivers, streams and reservoirs. This happens because industrial estate lacks waste water treatment systems, leading to all over pollution of water and contamination of quality of water (U.S.-A.E.P, 1996). The dying industry of Ludhiana (hub of woollen manufacturing industry) is releasing its industrial waste into the 'Budda Nala' and the River Sutlej, thereby contaminating the river water used for agricultural and household consumption. The water of the 'Budda Nala' is as contaminated with industrial effluents that its colour has turned to blackish and is unfit for even irrigation of the fields. Also, the sewage water is finding its way of the Budda Nala and as a result the micro-organisms count is too high. Micro-organisms found in human and animal wastes, such as, bacteria, viruses and protozoa, are the causes of many water borne diseases. These are virtually present in all the waste discharges, making it essential to treat drinking water. When phosphorus and nitrogen are drained through detergents and waste-run-off from agricultural fields; it enhances the growth of algae. When these two elements are discharged in water, they speed up the process called 'Eutrophication', which implies decline in oxygen content of the water. This factor explains that why Indian population largely suffers from problem of suffocation. At present 'Eutrophication' is resulting in decline and deterioration of water quality at all levels of all continents. In case of groundwater, nitrates from fertilizers and human and stock-wastes pollute it. It decreases 'Oxygen-Carrying' capacity of hemoglobin in blood, leading to Infant Mortality irrespective of gender discrimination (U.N. Environmental Project, 2000).

(iii) Solid-Waste Management

Solid wastes being thrown or disposed of by household, industries and agriculture in India, pose innumerable challenges which threat the Urban Environment along with its management. Though the need to improve solid-waste management techniques has been realized, still this part of urban environment management is severely strained. As per estimate, about 100,000 metric tons of municipal waste is generated daily in India. Per capita waste generation ranges from 200 gm to 600 gm per day. The collection efficiency is 70 to 90 per cent in metro-cities and 50 per cent in smaller cities. The Urban Local Bodies (ULBs) spend about Rs. 500 to 1,500 per ton for solid waste management. Out of this expenditure, 60-70 per cent is spent on street sweeping and collection, 20-30 per cent on transportation and even less than 5 per cent on final disposal (HUDCO and HABITAT, 2001). The management mechanisms for solid waste in cities, have not been able to attain a proportionate pace with that of disposal average of solid waste. In India, hazardous waste output is increasing due to chemical, textile and other industries (Sumathi, 1993; Frank, 1984; CEE, 1993). E-waste is also increasing at a pace that poses threat to human life through the radiation effect. Thus solid waste management is another serious problem that has not been tackled through an efficacious strategy.

(iv) Urban Environment and Transport: A Scene of Pollution

Transportation is an integral part of increasing population. Where human settlements and housing increase, the need of accessibility to these establishments also gains a momentum. Transportation is meant to have an access to these establishments. In developed and also in developing nation like India, many people prefer having their personalized transport, but there is an administrative shift towards public transport. Public transport is preferred, as this is the only way to effectively deal with energy crisis, congestion and other environmental concerns. It can carry a large number of people over longer distances using lesser road space and fuel per person and at the same time causes less pollution (Kumar, 2001). Both diesel and gasoline fuelled vehicles, specially two stroke gasoline engines used in two-wheelers and three-wheelers, emit CO_2. Hydro Carbon emissions from these vehicles are quite high, because a large part of the air-fuel mixture escapes unburnt into exhaust. So about 20-25 per cent of the fuel used is lost this way. This incidence is of severe concern in India; where near about 75-85 per cent of the registered vehicles are two and three-wheelers. Air pollution in urban areas is associated with higher incidence of respiratory ailments such as, bronchitis and asthma. Alarming ramifications of transportation problems can be well analyzed in case of urban areas. In cities, more number of private vehicles are witnessed which causes excess air as well as noise pollution. This is how transportation adversely affects the environment of a city. Urban Environment can be largely improved, if the urban transport is modified which is possible only if a strict pollution control check is made.. over the vehicles running in the cities.

(v) Health Sector: A Crucial Component of Urban Infrastructure

In the preceding part of the paper, infrastructural and environmental aspect of urban areas have been discussed. It can be well induced that deterioration in environment obviously leads to adverse impacts on health of individuals. In this context, administration needs to put emphasis on improvement of health and over-all well-being of individuals. A cross-nation analysis proves that as countries get richer, they spend more of their income on health care (World Bank Report, 1993). Indian economy has grown at around 6 per cent during the post-reforms period, but per capita Government spending on health has actually declined in the same span. This fact presents a satire to Indian situation, that health sector still needs to be made a priority, which has not yet happened. Table 9.4 shows a broad framework (of health expenditure, services and use, from 1980-1994) of various nations of the world divided on low, middle and high income basis and on the other hand, it includes statistical presentation of SAARC nations also.

Table 9.4: Health Expenditure, Services and Use

Name of the Economy/Country	*Health Expenditure*			*Physician (per 1000 pop.)*			*Hospital Beds (per 1000 pop.)*
	Public % of GDP 1990-95	*Private % of GDP 1990-95*	*Total % of GDP 1990-95*				
				1980	*1994*	*1980*	*1994*
World	3.2	2.8	5.4	1.0	1.4	3.5	3.8
Low Income	1.5	2.7	4.2	0.6	1.0	1.5	1.6
Middle Income	4.3	2.4	5.1	1.4	1.6	—	4.6
High Income	6.9	3.7	9.6	1.8	2.5	—	7.4
SAARC Countries	1.2	1.3	2.4	0.1	0.2	0.2	0.3
Bangladesh							
India	0.7	4.4	5.6	0.4	0.4	0.8	0.8
Nepal	1.2	3.8	5.0	0.0	0.1	0.2	0.2
Pakistan	0.8	2.7	3.5	0.3	0.5	0.6	0.7
Sri Lanka	1.4	0.4	1.9	0.1	0.1	2.9	2.7

Source: World Bank, World Development Indicators, 1998, Washington D.C., U.S.A.

It represents the percentage of both public and private expenditure on health. Indian data is presented under SAARC category, but virtually the Indian statistic does not show a satisfactory scene. It even falls below the level of low income nations of the world; at some instances. Despite development in other aspects, health sector of India has not been to achieve the expected goal of development.

Both the health provisions and financing are considered to be a State's subject. It can be incurred that the State's share (on an average) on health spending, ranges about 80 per cent of its total expenditure, whereas, centre fully finance Family Welfare Programmes. Most of national disease programmes are funded on 50:50 share arrangements with States. Out of total expenditure on medicine, education and research, the Central Governments share little over 40 per cent only. States bear almost all the expenditure. Hence, State's economic condition and human and financial resources have a direct bearing on health outcomes. This has been witnessed in post-reforms period, that there is a close link between per capita State domestic product, per capita State health expenditure and the health status. Unfortunately, economic reforms in India since 1990 have rarely been accompanied by health sector's improvement. The central budgetary allocation to health sector is being presented in the following part of the paper, which proves that health sectors improvements have not been brought up; as it was required. In the years 1991-92, 1.2 per cent was allocated to health services, 1.4 per cent in 1993-94 to 1998-99; 1.7 per cent in 1999-2000 to 2000-01 and 1.8 per cent in 2001-02. It is calculated that the allocation made to the health sector (of the total India's budget) range from Rs. 1,734 crores in 1991-92 to Rs. 6,752 crores in 2001-02. Though the investment made on health services are convincing, even then the results are not as remarkable, as they ought to be. The reason behind ineffectivity of these allocations is the absence of State based efforts. Environment of cities (in context of health) can't be improved till States participate. High income earning families can afford the expenses of treatment, but middle-class or poor families fail to avail treatment in hospitals and costly clinics which are after located in metropolitan or big cities. Even in case of Government hospitals, there is lack of manpower, commitment to serve, equipments and their maintenance, medicine and related facilities, which again reinforce the inability of the State Government in regulation and control over health sector's institutions. This is the factual scene of health services in Indian urban areas. It needs a capable urban environment management so as to check atleast 50 per cent inflictions on health. This requires commitment to the health sector both on administrative and service level by those associated with the management of urban affairs.

SECTION II

Importance of Community Participation with a special Reference to NGOs and CBOs and the Role of Urban Local Bodies (ULBs)

Local level participation is an effective criteria mandatory for environmental protection and maintenance. Urban Local Bodies (ULBs), NGOs and Community Based Organizations (CBOs) need to be empowered through decentralized paradigm of governance. The 74th Amendment made in Indian Constitution laid important services to be performed by ULBs, to combat with environmental degradation. A number of functions listed in the Twelfth Schedule to Indian

Constitution, are construed as 'Environment Services'. These include regulations of land use and building construction, urban health, sanitation, waste management and conservancy, urban forestry, promotion of ecological aspects, urban amenities like parks, gardens, management of electric crematoriums, regulation of slaughter houses and tanneries, etc. Many other provisions were also provided to ULBs. To perform all these functions, it is necessary to mobilize the capital funds. For instance, 'urban forestry' is not an easy target to be achieved, specially in an era when each kilometre is dominated by some or other form of human settlement with no space for forestry.

At this instance, it is important to analyze the revenue expenditure and revenue gaps of municipalities of States, where a well planned environmental management is to be initiated. Table 9.5, presents a profile of revenue, expenditure and revenue gaps of municipalities of some States of India, in the year 1993-94 and 1997-98.

Table 9.5: Revenue, Expenditure and Revenue Gaps of Municipalities in Some States 1993-94 and 1997-98

States	*1993-94*			*1997-98*		
	Total Revenue	*Total Expen-diture*	*Revenue Gap*	*Total Revenue*	*Total Expen-diture*	*Revenue Gap*
Andhra Pradesh	193.94	171.33	22.61	321.55	291.88	29.67
Gujarat	181.45	191.05	-10.05	341.11	369.61	-28.05
Karnataka	110.23	93.75	16.48	137.21	131.24	5.97
Kerala	72.22	85.12	-12.09	175.19	177.98	-2.79
Madhya Pradesh	62.92	70.27	-7.35	113.03	140.08	-27.77
Orissa	20.72	53.08	-33.08	36.79	62.42	-25.63
Punjab	126.03	136.54	-10.24	221.96	230.67	-8.71
Rajasthan	77.92	52.04	25.88	160.22	123.39	36.83
Tamil Nadu	247.33	200.48	46.85	630.03	393.63	236.04

Source : Report of the Eleventh Finance Commission, 2000-05, Annexure vii 3D.

Government of India

The report submitted by the Finance Commission shows that in case of States like Gujarat, Madhya Pradesh and Orissa a severe revenue gap exists as compared to other States. On the other hand, States like Andhra Pradesh, Karnataka, Rajasthan, Tamil Nadu show a better picture. Anyhow, more funds need to be generated if the condition of these States is to be improved. Financial capacity-building of ULBs must be aimed at, which is possible only if local functions are entrusted to the ULBs and law restricts the power of the State Governments to remove such function from the domain of municipalities (Ghosh, 2003), improving physical and social facilities.

Role of NGOs/CBOs/Communities in Maintenance of Urban Environment

'Civil Society' is a very popular term prominent in modern debates. 'Civil Society' denotes a broader spectrum which includes Non-Governmental Organizations (NGOs), Community Based Organizations (CBOs), Citizen's Group, professional organization, etc. Many NGOs are functioning in India and are involved in various motives like economic development (building rounds, embankments); housing and environment (in rural as well as urban environment), social development (health education), cultural development (artistic and literary), etc. In Urban Environment Management, NGOs and CBOs must work in partnership with International agencies and national organizations, urban local authorities, elected officials, individual households, common citizens and media. Many programmes initiated by United Nations, make it obligatory for NGOs and CBOs to participate in designing and implementation of the programme. Some experiences are being discussed in the following part of the paper.

- Kitagata, in Japan, in (1980) initiated an Environmental Protection Programme. In this region backward community called 'Bura Kumins' reside, which faced a severe environmental degradation. This project was assisted by Wakataka Planning Research Institute, and resulted in a great success as it improved various environmental aspects. It succeeded in plans related to solid waste management and increasing the green cover.
- 'EXNORA' a local NGO in 'Chennai' is also an imminent work which aimed at solid waste management since 1989. It attained financial assistance from UNSAID. EXNORA collects 600 metric tons (out of city's 3000 tons) of Solid

Waste. It proves that NGOs, with collective efforts of common people can pave a way for multiple improvements.

- In Pakistan, Sindh Katchi Abadi Autonomy (SKAA), provides help in Solid Waste Management through small-scale communities. The Municipal Corporation organizes secondary collection and disposal.
- 'Sevantha' is a well known NGO in Sri Lanka. It also successfully deals with sewerage networks. It works on self-paid basis. It has also been able to achieve a great success in environmental protection in Sri Lanka.
- Burigaya Bachao Andolan—Another functioning NGO in Bangladesh, is functioning with the help of 25 organization and educational institutions. It was initiated on 21 July 2001. It aims at protection of 'Burigaya River' which suffers from environmental degradation.
- Some other examples are:
  Bangladesh Poribesh Andolan (BAPA) and Poribesh Rokkha Sopoth (POROSH), both working for environmental protection in Bangladesh.
- Sustainable Environmental Management Programme (SEMP), supported by UNDP, is a recent example of Bangladesh. UNDP's 'Local Initiatives for Environment' (LIFE) is also in exemplary projection of NGO's and CBO's role in environmental improvement. Some NGOs bring forth their own projects and work for their self-determined motives without any super ordinate assistance.
- Sustainable Implementation of National Plans and Actions (SINPA) is another programme run by an NGO in Bangladesh to implement HABITAT agenda. It is funded by Netherlands Development Corporation (NDC). It offers services of Institute of Housing Settlement (IRS) to Bangladesh, Bolivia and Zambia. Its local partner is 'RADOL'—a local NGO working for environmental causes like solid waste management; traffic management; slums improvement; housing for poor (Ministry of Housing and Public Works, Government of Bangladesh, 2001).
- ORANGI PILOT PROJECT of Karachi is another milestone, which was initiated in an unauthorized settlement, covering over 8,000 hectares, with around 1 million inhabitants. People

built their own homes themselves as they did not have any official help. This area did not have any institutionalized sewerage system, so the drainage pipes used to remain filled with dirt and the houses constructed on both the sides of these drainage pipes, would sting. This NGO project united hands with local communities and they build their own destiny. This project organized meetings of 10-15 houses at a time and made them aware of negative implication of absence of sewerage system. This effect was a great success as it could make 69,000 sanitary pour-flush latrines in homes, 4,459 sewerage lines and 345 secondary drains (UNCHS, 1996).

- Kempung Improvement Programme (KIP) of Indonesia is another example. Jakarta and Surabaya were awarded with 'Aga Khan Award in Architecture' for urban management and improvement. Surabaya's Municipal Authority, affected the local economy and undermined its ecological sustainability and the new investments did not give adequate returns.
- In Andhra Pradesh, Government has initiated a path breaking strategy to make Hyderabad and other cities of the State, green and clean. This city was known for its gardens, orchids and water-bodies, and had more than 500 water bodies, which now constitute only 150. It still has five major parks and 240 small parks. A poor infrastructural network and severe resource depletion adversely effected the local economy and undermined its ecological sustainability and the new investments did not give adequate returns. A common population based project named 'JANAMBHOOMI' was initiated in January 1997 and 'CLEAN AND GREEN' in August 1998. Both of these initiatives included public efforts. Solid waste collection, which earlier was not more than 20 per cent, has shoot up to 85-90 per cent. Municipal Corporation of Hyderabad is now regularly taking up tasks like widening of roads, removal of illegal construction, lifting garbage, public toilets and their cleanliness. This public and private partnership laid an example and the State stood first in improvement of green cover with its share occupying 965 of the 3,000 sq. km. in the country.

Rationale behind discussing these examples is to interpret the real worth of NGOs, CBOs, and Local Authorities in management of urban environment.

Conclusion

Problems, challenges and their solutions emerge from same root. When excess population poses combatable issues, it also proposes hands for their solution. But there is a need to realize the hidden assisting capacity and involve the common public in the improvement process. If urban environment is getting deteriorated in context of sewerage, health, air and water pollution, solid waste management, it needs to be checked immediately. Only Government or Municipal Corporation can not solve the problem. Any initiative, project or programme needs to be invested by common people, MCs, State Government and Central Governments together, because environment belongs to individuals and the owners are liable for the maintenance of their possessions, whereas the State must provide support to the public projects. The 12th Finance Commission has made a series of major fiscal recommendation, including issue of the share of States from divisible pool of resources to 30.5 per cent from the exiting 29.5 per cent and called for doing away with the present system of Central Assistance to the State plans in the form of grants and loans. It has recommended a State transfer to States of Rs. 7,55,752 crores (*Tribune,* 2005). This recommendation will hopefully be a boon for development of Indian States followed by cities, towns and villages. But still evaluation of success of a geological unit is not efficacious, until the assessment instrument involves all units in the development process. Environment Assessment (EA) has become important because it has shifted from a single dimensional approach oriented to economic vision solely, to a multifaceted approach. EA aims at Risk Analysis which is helpful to achieve the desirable and a justifiable format of growth and development. EA needs to be made widely applied in all the aspects of Urban Environment Management. Concluding, urban environment needs to be tackled through a strategic planning and zealous endeavours of the Government and the communities involved in the process.

REFERENCES

Census of India, 2001.

Centre for Environment Education (CEE), "Study of city circle; Best Practices in Solid Waste Management and Collection, Transportation and Disposal", *CEE,* Bangalore, 1993.

Cranney Brenda, *Local Environment and lived Experience—The Mountain Women of Himachal Pradesh,* Sage Publications, New Delhi, 2001.

Frank Flintoff, "Management of solid waste in developing countries", *World Health Organization,* Regional Office of South-East Area, New Delhi, 1989.

Ginkle H.Vans, Barrette B., Julius C. and Velasquez, *Human Development and the Environment; Challenges for the United Nation in New Millennium,* Rawat Publications, New Delhi, 2003.

Ghosh A., *Urban Environment Management Local Government and Community Action,* Institute of Social Sciences and Fredrich Ebert Stiflung, Concept Publishing Company, New Delhi, 2003.

Housing and Urban Development Corporation (HUDCO) and United Nations Centre for Human Settlement (HABITAT), *The State of Indian Cities,* Nairobai, New Delhi, 2001.

Jayaplalan N., *Urban Sociology,* Atlantic Publishers and Distributors, New Delhi, 2002.

Kumar H.D. *Sustainable Human Ecology,* Raj Kamal Electric Press, New Delhi, 2001.

Ministry of Finance, *Housing and Public Work,* Bangladesh National Report, Government of Bangladesh, 2001.

Panda Samiran, *Living with AIDS virus—the Epidemic and the Responses in India,* Sage Publications, New Delhi, 2002.

Reddy, Narayana G., *Building-capacity for Sustainable Development,* Kanishka Publisher, New Delhi, 2002.

Sumathi S., *Solid Waste Management: Best Practices adapted in Madras city,* Centre for Environment Education [CEE], Bangalore, 1993.

Tribune (2005), February 27th, 2005, (p. 21).

United Nation Corporation on Housing and Settlement (1996), *An Urbanizing World-Global People on Human Settlement,* Oxford Press, London, 1996.

United Nations Environment Project, *Global Environment Outlook,* Earthscan, London, 2000.

United States—Asia Environment Partnership (U.S-A.E.P.), *Environment Infrastructure Strategy in India,* Washington D.C., 2001.

World Development Report (1999).

World Development Report, *Sustainable Development in a Dynamic World—Transforming Institution, Growth and Clarity of Life,* Oxford University Press, London, 2003.

10

Environmental Degeneration with Focused Reference to Deforestation and its Impact on Developmental Sustainability

JASLEEN KEWLANI

Introduction

The greatest crisis, the whole globe is facing at present is 'Environmental Degradation'. It is affecting the geology of the earth as well as the communities dwelling in it. The increasing population of the world puts forth, proportionately increasing demands to maintain a glib survival. Fulfilment of these demands requires an efficient developmental process. Development is eternal, if it is sustainable. Lack of 'sustainability' in developmental process leads to alarming consequences, since quest for short-term benefits for the present generation, results in an uncontrolled exploitation of the resources. It virtually proves infructuous for further development and puts the future generation at a risk. Sustainable development incorporates improvement and maintenance of infrastructure, land and soil, flora as well as fauna and occupational structure in such a manner that it leads to economic growth, without sacrificing environmental quality. In a country like India with a high population density and high level of poverty, every ecological niche is occupied by some occupational or cultural group for its sustenance. If that ecological niche is destroyed, or its resources are unevenly allocated (more to the powerful and lesser to the weaker groups) it will have an incommensurable effect (Aggarwal, 1996). In such a situation the dispossessed groups will certainly suffer. This adverse effect is not limited to a specific area only; rather it catches communities from the globe as its preys. If this phenomenon is not controlled, environmental 'Myopia' would appear to be a societal disease (Roy and Aggarwal, 1996).

Objectives

Interpreting the importance of the issue of environmental degradation, the paper aims at discussing the following objectives:

- Deforestation and Economic Development;
- Impact of Deforestation on Various Communities;
- Sustainable Development.

The methodology applied for collecting information includes analysis of secondary sources of information. Various books on environmental issues, magazines, containing allied topics have been used to accomplish the needed information. Other than this, some important documents, journals have also been consulted.

Deforestation and Economic Development

Though, there has been a civic-cum-administrative consideration for environmental protection and even the framers of Constitution of India and law-makers have given a significant space to provisions of adjudication in this context, but still a paradoxical situation exists. There is probably no other area of India's geological domain, that has been more viciously attacked and destroyed in the last century than country's environment, specifically the forests. A report of National Committee on Environmental Planning in the last decade clearly stated that no more than about 12 per cent of country's land surface is under adequate tree cover. An Aerial Survey conducted by the Indian Space Research Organization (1974) in Anantpur District of Andhra Pradesh found that many Reserve Forests in the District had no trees at all. There is not a single cause behind deforestation or other forms of environmental degradation, but a complex matrix of inter-knit and inter-webbed factors acts as a pushing force. These factors are river valley projects, agricultural expansion projects, roads and transportation development, industrial expansion, etc. The reason may be any, but this form of environmental degradation poses severe threats to the glib survival of humans. Forests ecosystems do not only include forests, but also climate, soil, atmosphere and all living species. A reduction in green layer or green cover leads to devastating loss to flora, fauna, and mankind and even anticipates losses for the future generation also (Shafi and Raza, 1992). Having indepth gauge into the issue, it is inferred that economic forces are at the back of environmental degradation. There exists a strong link between enlarged disposition of natural resources of natural environment and mastery of nature through scientific knowledge which is popularly known as technology (Bhattacharya, 2001). Where technological development provides multiple advantages to mankind, it demands some sacrifices also, causing irreparable loss to the human survival. This form of development moves ahead by rampantly pushing back the frontiers of nature.

There is probably no other area of India's environment that has been more viciously attacked and destroyed in the last century, than the country's forests. Prime Minister of India, Late Smt. Indira Gandhi, while addressing Central Forestry Board in 1981, frankly admitted that only half of the official forests lands are under adequate tree cover. There is not a single cause behind this uncontrolled deforestation, rather multiple factors work at the back of phenomenon. The following table shows the forest area lost between 1951 and 1972 and causes for this loss.

Table 10.1: Forest Area Lost Between 1951 and 1972

Purpose	*Area (in Thousand Ha)*
River Valley Projects	401
Agricultural Purposes	2433
Roads and Communication	55
Establishment of Industries	125
Miscellaneous	388
Total Area Lost	3402

Note: This represents an average annual loss of 155,000 ha.

Source: Forest Resource of Tropical Asia, FAO, 1981.

The data provided in the table indicates that a large amount of forest land has been converted into agricultural land for agricultural purposes followed by other causes. The new economic policies and the structural adjustment programme often predict environmental impact without accuracy because of the complex interplay of various economic, social, political and ecological factors. These are put into effect in an effort to meet India's severe balance of payments crisis, and to propel its economy into quicker growth and global integration. A part from direct fiscal policies, the major components of the new package include boosting exports to earn foreign exchange, liberalizing industrial production, dropping barriers to the entry of foreign companies and goods expanding privatization and cutting Government spending.

Indian companies are also squeezing natural resources at a faster pace for shrimp farming in most valuable brackish wetlands and lakes. Agri-business consortium has been proposed by the Government, ostensibly to help small and marginal farmers, which is to give a boost to commercializing Indian agriculture to cater to expanding consumerist markets. In the new liberalized atmosphere, the environment departments in all States and at the Centre are going to become everyone's punching bag and will find it harder and harder to enforce their regulations. Industries have been cleared adjacent to a Marine National Park and protected areas of wildlife officials. The destructive potential of the liberalized economy justifiably identified tropical islands, rainforests, coral reefs and marine waters and genetic store houses. World Summit for Sustainable Development in Johannesburg in its report entitled "Global Challenge Global Opportunity" has identified water and sanitation, energy, agricultural productivity, bio-diversity, and human health, as the moot point. UN Environment Programme (UNEP) believed that "A consumer lifestyle based on the desire for mobility, convenience and product disposability has undercut the further advancement of resource efficiency and waste reduction". South Africa President Thabo Mbeki that "a global human society based on poverty for many and prosperity for a few, characterized by islands of wealth surrounded by a sea of poverty, is unsustainable. The tragic result of this was the avoidable increase in human misery and ecological degradation, as seen in the growth of the gap between North and South. The international community needs to fight the global apartheid; the elimination of the global apartheid between rich and poor."

The reasons may be any, but deforestation poses severe threats to survival of humans. Forests ecosystems do not only include forests, climate, soil, atmosphere rather, all living species are also integral parts of this entity, which clearly states and predicts an anticipation of a bombardic loss at present and in future also (Shaffi, Raiza, 1992). Realizing an important and solid linkage between economy and environment, many statements can be laid. One of such statements is that it has been observed that the forces which are at the back of environmental degradation, are purely or partially economic in nature (Bhattacharya, 2001) posing a threat to sustainability. There exist a link between the enlarged disposition over the resources of natural environment and mastery of nature through scientific knowledge. This link has been described as technology. Where technology provides enumerable advantages to life, it also demands some sacrifices, which cause irreparable loss to mankind. Technological development moves ahead by rampantly pushing back frontiers of nature. There are many factors behind deforestation. Population and its density is increasing on a fast pace. This accelerated density demands adequate housing facilities. To cope with this demand, when enough area of the city land has already been turned to housing area, forests become the target. Forests are cut down to facilitate public in context of housing; in fact depriving people of fresh air, forest product and green cover.

Economy can be developed by establishing more and more industries, which in turn needs a sufficient power generation. Both phenomenons destroy forests. Industries replace forests; large establishments are made in these environmental entities, which aims at paving a way for economic development. Consequently a great demand for power generation emerges. For the purpose of providing power and electricity to industries, to make them function successfully, dams are constructed. Where dams have got a great potential to control floods, construction of several large dams involve the eviction of thousands of families, and their resettlement at some other site. One of the first major eviction operations related to the town of Bilaspur in Himachal Pradesh which was surveyed by the Bhakra Dam Reservoir. Seven thousand citizens of the town, with modern amenities like a college, a hospital, markets and playgrounds, were displaced

(*Ibid.*, p. 63). The devastating impact of dam construction does not end here. This civil engineering task leads to harmful ecological impacts. The environment of the catchments regions gets wantonly exploited. The decimation of vital forests in the catchments areas is the most serious consequence of this engineering approach. The immediate victims of this ecological decline are the tribal, the peasants in the river valleys and the folk, whose life depends on the resources provided by the nature and environment around them. Adding to severity of the effects of dam construction, rapid deforestation in the catchments area leads to increased soil erosion, landslides and consequently heavier water-flows into the reservoirs, imperils river valley projects in Himalayas, Western Ghats and Chotanagpur regions.

Construction of dams for economic motives is not solely responsible for uncontrolled deforestation; rather its results again become a cause for further intencification of deforestation. Forests are cut to resettle and rehabilitate evacuees as well as to provide residence and transportation for the employees of the sites. This rather implies 'planned destruction of forests' for the purpose of economic development.

The most emphatically featured element of economic development is its hegemonic impact on the tribal, and rural folk who depend upon forests even for their ordinary survival to a large extent. These communities reside in proximity to the forests and hence get largely adversely destroyed when deforestation occurs. Deforestation leads to increased soil erosion, landslides and consequently heavier water flows into reservoirs and hence, imperils river valley projects in Himalayas, Western Ghats and Chotanagpur region. Hence, commercial exploitation of natural resources has come out to be the most unfortunate set back to the forests. Not only does nature place limits on societal development, not only does societal development effect nature, but the effects of societal development on nature rebound on human societies again (Martell, 1994).

Tribal and rural people are largely dependent upon forests for either fuel and fire needs. Rural and tribal populations exploit forest and consider these exploitation as their basic right, being emotionally and possessively attached to these forest owned by them. They cut and use forest wood for various purposes (domestic as well as non-domestic). It makes it difficult for the Government to make a correct assessment of fuel requirement of the national population on two bases. Firstly, when rural communities get fuel from forests as their right and it goes unrecorded and secondly, pilferage of fuel wood from forests also goes unrecorded. These are the challenges before Government and public, which combat the smooth survival pattern, if deforestation is done in an uncontrolled manner.

United Nations Environmental Project (UNEP) elaborates a fact that the global warming is a highly contravened issue and is the most stimulating cause of Asia Haze. UNEP blamed developing Asian Nations for the 'Brown Haze' being formed on Indian Ocean and it is stated to be alarming for global atmospheres. According to its report, the first brown haze was an effect of Indonesian forest fires of 1999; Forests fires, biomass burning and other pollutants in Asia are threatening the earth's environment (Singh, 2000). The process of deforestation renders pernicious and devastating influences which widely include extinction of species; erosion of soil resulting into accelerating silitation in dams rivers and the coastal zones; global warming; loss of fuel, food and mineral; and above all the loss of culture and way of life because many tribal and rural community identify their survivalhood with forests and the trees thereon. In all, real situation of Indian environment is worst and one-dimensional approach of economic activities will ultimately lead to an absolute environmental disaster (Kayasthas, 1992).

Impact of Deforestation on Various Communities

Forests are the protective and productive entities of environment. A study of Kumaon area refers to the importance of forests and states that these are a crucial source of fuel wood and fodder for people of Dharamgarh Village. With process of deforestation, the worst sufferers ought to be the women of these dependent areas, whose main duty is collection of wood and fodder. As a result of deforestation, they have to spend more time and have to move much far from their residential areas to collect fuel wood and fodder. In such a situation, it is the girl child who is withdrawn from her school to provide a hand to mother's household activities, work depriving her of basic right to education. Not only the intellect, but the health of females also suffers a lot of loss as they have to cope with this increasing burden with their limited physical capacity and work potential (Aggarwal, 1996).

Discussing the case of tribal, the condition has worsened. Problems encountered and tolerated by them are rooted into the process of growth and development paradigm, which is a great paradox. Development of economy aimed by States, has led to further pauperization of tribal people. Deforestation has disturbed the original life of the interior-forest-village dwelling tribal. This process has provided some facilities to tribal like transportation, schools, roads, health intrastructure, but the *vice versa* aspect is that these benefits in real deprive them of their right to have a healthy environment around them (Punalikar, 2000).

Non-conservation of bio-diversity affects not only human communities but other species too. Bio-diversity is not a vacuumed concept rather it includes species, ecosystems and genetic diversity also. Deforestation degrades environmental standards, which further makes exclusive species extinct. It is indeed the most alarming consequence, since species are such biological entities, which cannot be recreated by men, if once lost (Haaningam, 1997). These species are wealth of nation as they are a boon for entertainment and cater to economic motives also, which fetch good amount of revenue from zoos and sanctuaries. Deforestation has resulted in extinction of 15-20 per cent of species by the end of this century. According to an estimate, about 0.2 to 0.3 per cent of all the species in forests are lost every year and there is a loss of one species everyday. The effects of deforestation on dependent beings are so intense, that the whole living scenario gets affected. The societies like rural, tribal, which developed deep roots in their soil, incur irreparable damages as forests provides life protecting medicines, deities, demons and totems to construct a religion of their own (Chauhan, 1998; Aggarwal, 2002; Vyas, 1996; Sinha, 1977). Environmental degradation in the form of deforestation has caused devastating effects on the whole world's geology as well as humanity. This practice needs to be controlled immediately as it may provide a fast and short-term benefit to a section of the present generation, but deprives the offshoot of it, of its right to develop. Deforestation anticipates cious and ecological panic for the future generation also because the need of trees will remain intact, but there will be no tree for them, if deforestation continues at the same pace and pattern, as it is being carried at present. In technical or expertise term, deforestation needs immediate check over it, if a sustained growth and development of nation is intended.

Sustainable Development

The term 'Sustainable Development' was coined at the Cocoyoc Declaration on Environment and Development in 1970. It is such a form of development which was advocated for, in many events like Stockholm Declaration (1972); Brundtand Commission Report (1987) etc. Sustainability is a crucial criterion for development as we can not sacrifice the future, to make the present contended. Sustainable development is development that meets the needs of the present, without compromising the ability of future generations to meet their own needs. To bring about such a sustained society, the development paradigm needs to be economically viable and environment friendly as well. Evanescent or short-term benefits often turn to failures in long-term. To attain long term and eternal advantages, exploitation of resources, direction of investments, orientation of technological development and institutional changes must be consistent with the requirements of the present and future both. A symboitic relationship between consumer human race and producer natural system needs to be recognized, which implies compatibility between economy and ecology (Khanna, 2000). The sustained pattern of development is an impossible target, unless and until various nations co-operate each other. The reason behind this plea is that human race can be demarcated or divided on the basis of geographical boundaries anyhow, but an interaction cannot be resisted in case of global atmosphere or environment. Paradoxically, the lack of a synonimity lessens the possibility of this international co-operation. 'Pros' of one nation become 'cons' of the other, over the same issue, at same point of time. For example, rainforests are of a great concern to whole mankind, but for nation like Malaysia, rainforests are a source of economic upliftment and development. The economic resource development in Malaysia is brought by exploiting rainforests. Global pressure has been there, but Malaysia has put forward notified causes for the rainforest exploitation. This is a situation of conflicting interests, which is a powerful hindrance in the way of Sustainable Development.

Regular debates on environment have increased environmental awareness among many nations in the past few decades producing some fruitful effects. Many international agreements have been initiated. Where a nation is unable to formally become a part of such a unanimous deal, unilateral steps are being taken. Among these agreements, some important examples are being cited below:

- CITES: Convention on International Trade in Endangered Species of Wild Fauna and Flora, adopted on 3rd March 1973 and enacted on 1st July 1975.
- BASEL: Convention on Control of Transboundary Wastes and their Disposal, adopted on 22nd March 1989 and came into force on 5th May 1992.
- 'MONTREAL PROTOCOL' on depletion of Ozone Layer (1987), etc.

Such efforts can make Sustainable Development, an aim in proximity. Until and unless the local environmental interests are matched with that of the international ones, environment management at a longer and regular pace is impossible.

Resilience

A new interpretation to sustainability concept through the use of word, 'Resilience', is also an efficient instrument, which implies an ability and power of a system to face perturbation, shocks and surprises being incurred upon it. This is infact the capacity of a system to absorb disturbances without undergoing any fundamental changes. The term 'resilience' provides more practical interpretation of the concept of sustainability. It can be simply analyzed that Sustainable Development does not only mean simply 'having' ability, rather 'maintaining' the ability to retain its fundamental credentials, so that it can be used by the future generation also.

Effective Mechanisms for Sustainability

Environmental degradation in form of deforestation, water depletion, drought, soil erosion, etc., has caught the whole mankind into its prey. The problem has transcended local as well as national boundaries and reached an International pedestal now. Hence, it will be justified if 'epidemic' term is used to refer to environmental deterioration. It is affecting every individual, all species and the geology of the world, which in fact reinforces intense scarcity of natural resources. Solution of this issue needs to involve all communities, governmental encumbents, institutions, non-state actors and common masses too. In the following part of the paper, we will discuss some possibly practical solutions to curb the problem of Environmental Degradation. Some of these steps have already been initiated, in context of which strict implementation needs to be made, so as to bring about the expected changes.

To check unregulated use of environmental resources, the application and implementation of Environmental Impact Assessment (EIA) programme is an important concern. Major feature of this programme is that it involves search for the methods to manage environment in compatibility with the complex world of today. Such an attempt will help save the nature from ever increasing materialistic and exploitative tendencies of men and defend the environment from destructive and inadvertent activities of human beings to develop the economy (Kumar, 2002).

Another helpful instrument which can pave way for immense environmental improvement and modification is 'Environment Cost Benefits Analysis', which is used when a purposive project is launched. Traditionally, this cost benefit assessment used to be expressed only in economic terms, such as in terms of labour interests, borrowed capital, etc. But all this has not produced those desired results, which the community actually expected. If Sustainable Development is to be brought, then the costs of benefits associated with the projects and environmental impacts must be taken into account. This can make it convenient to predict any environmental changes expected to be incurred as a consequence of operating a project, in lieu of which a pre-hand precaution for an efficient management can be taken. This approach will be helpful to take up qualitative ramifications.

Community participation is the most efficacious factor having potential to function in a productive manner in environmental management. But this participation must go on systematically not randomly. It must bifurcate the contributions into two parts; first are those mainly amenable to community management and the rest to global management of environment (Aggarwal, 2002). One of such successful step was; United States National Environmental Policy Acts (NEPA) 1969. It aimed at involving the Government, industry, community at large, but India has never used it as a milestone in the management paradigm. The results can be the most desirous and expected ones if the public which suffers, plays a large foot of role in addressal of its own grievances. This needs a moral boost up by the Government.

Joint Forest Management (JFM) is another corporate activity, considered to be usable and potential in Environment Management. It needs a real sharing of responsibility and decision-making between the States and the local communities (Chahal, 1994). Whose possessed entities, i.e. forests have been snatched from them. The communities dependent upon the forests or those who have lost forests around them due to excess deforestation need to put forwards contributive hand to bring about an improvement, by managing forests, in collaboration with the government.

Environmental ethics and an 'ism' work at the back of community participation. 'Ism' implies transformation of ethics into behavioural aspect. This transformation can be carried forward through the enlightened groups, which are often called Non-governmental Organization (NGOs). These NGOs put resources together and make the required resources available in case of scarcity, and move ahead to achieve the aimed target. NGOs work and make others work because they have an influencing leadership quality, which directionalize the efforts and thoughts of public towards managing Environment. Especially, in a nation like India with a huge population, agencies working as 'middle-man' are needed which effectively negotiate between the public from all corners of the nation and the government. The government must replenish its legislative and executive circuits. Paucity of strict laws as well strong implementations of environmental laws put whole community at a risk. Where cutting forests is a crime, there only, no strict check is maintained, which virtually facilitates common folks, industrialists and developers to make use of this leniency.

In order to strengthen the environmental management capacity at national, state and local level, the Ministry of Environment and Forest had launched an Environmental Management Capacity Building Technical Assistance Project in 1997 with assistance from the International Development Agency. This project has an outlay of Rs. 140.62 crores. The areas identified are capacity environmental administration, decentralization of environmental management, implementation of environmental laws, and monitoring compliance in specific priority areas (Economic Survey, 2004). At the same time, industrial growth has been at below 10 per cent for several years now. One of the most important challenges in Indian economic policy consists of devising strategies for obtaining industrial growth in excess of 10 per cent, which infact bestows deterioration to the national environment.

Last but not the least, Environmental Education must be spread. It must consider environment in its totality, viz. economic, political, social, technological, cultural, historical, moral and aesthetic aspects. A holistic approach helps bring about continuity in the process of learning. Certainly, the scope of Environmental Education and its success gets affected by factors like size and growth of population, pollution, human health and quality of life, etc., but it can be made a success if this education is related with environmental sensitivity, knowledge, problem-solving skill and value clarifications.

Conclusion

A mature analysis never separates economic and ecological issues. If any of these is ignored, then the result is not a sustained one. Outcomes gained from such an approach prove to be short-term benefits which do not provide long-term progress. No human being can survive without the essential services provided by the ecosystems like, land, air, water and forests. These services are generated by interactions among organisms and the physical as well as chemical environment, in which they live. These indispensable functions of ecosystems maintain a genetic library which sustains the process of preserving and regenerating soil, recycle nutrient, control floods, filter pollutants, assimilate wastes, pollinate crops, operate the hydrological cycle and maintain gaseous composition of the atmosphere. This genetic library acts as a 'Carbon Sink' and oxygen producer and perform many more such contributive functions. In such a situation forests are a panacea for humanity and their depletion is a curse. Environment degradation needs to be controlled by preserving ecosystems of the world, or it would lead to cious in survival and a blocked development pace. At least, it can be said that in developed, underdeveloped or developing nations, no account of economic development will be regarded as adequate, if environment resource base were absent from it. This implies that economic development and environment maintenance must be carried forward simultaneously, if sustainable development is aimed at in authentic terms.

REFERENCES

Aggarwal Anil and Narain Sunita, "*State of Indias Environment; the First Citizens, Report*"; Centre for Science and Environment, New Delhi, 1996.

Aggarwal S.K. and Dubey P.S., "*Environmental Controversies*", A.P.H. Publishing Corporation, New Delhi, 2002 (pp. 293-305).

Bhattacharya B., "Environment and Trade; Some Issues" in "*Environment Management and Andit*", Rao P. Sasi Bhushan and Rao R. Mohana, Deep and Deep Publications Pvt. Ltd., New Delhi, 2001.

Chahal, S.K., *Environment and the World Life : Towards a New Paradigm*, Ashish Publishing Company, New Delhi, 1994.

Chauhan I.S. and Arun, *Environmental Degradation; Socio-economic Consequences*, Rawat Publications, New Delhi, 1998.

Economic Survey, Government of India, a Ministry of Finance Publication, New Delhi, 2004.

Hanningam A. John, *Environmental Sociology: A Social Constructionist Perspective*, Routledge Publishers, London, 1997. (pp. 146-162).

Kayasthas S.L., "Environment and Development—Emerging Perspectives and Imperatives", in *Development and Sociology-Essays in Honour of Prof. Mahammad Shafi*, Rawat Publications, New Delhi, 1992 (pp. 39-67). See also Peter and Calvert Susan. "The South, the North and the Environment". Biddles Ltd., Kings Lynn and Guilford Press England, 1999.

Khanna P. and Basu Ram, "Environmental Evaluation of Economic Growth," in *Environmental Management and Audit*, by Rao P. Bhusana Sasi and Rao R. Mohana, Deep and Deep Publication Pvt. Ltd., New Delhi, 2002.

Kumar Arvind, *Environmental Pollution and Agriculture*, Efficient Offset Printers, New Delhi, 2002, (pp. 477-482).

Lee A. James, *The Environment, Public Health and Human Ecology; Consideration for Economic Development*, the Johns Hopkins University Press, London, 1985.

Martell Luke, *Ecology and Society—An Introduction,* Polity Press, Great Britain, 1994 (p. 23).

Patnaik Nabe Kumar and Rao P. Sasi Bhusana, "*Environmental Education for 21st Century,*" in *Environment Management and Audit* by Rao P. Sasi Bhusana and Rao Mohana, Deep and Deep Publication, New Delhi, 2000. (pp. 201-211).

Punalikar S.P., "Influences between Development and Environment; a micro-narative on tribal society in Gujarat", in *Resource Management and Contours of Development*, by Barik B.C. (2000).

Roy Dunn and Agarwal Anil, "*Politics of Environment in State of India's Environment: The Second Citizens Report*", by Anil Aggarwal and Sunita Narain, Centre for Science and Environment, New Delhi, 1996.

Shaffi Mohammad, Raza Mehndi, "*Forestry Ecosystem of the World,*" Nice Printers, New Delhi, 1992.

Singh Uma Shankar, "Fighting Forests Fires", in *Environment Management and Audit* by Rao P.Sasi Bhusana and Rao P. Mohana, Deep and Deep Publication Pvt. Ltd., New Delhi, 2000. See also Kellywich and Granvich Sarah "Global Warming and Development", in *People and Environment* by Morse Stephen and Stocking Michael, UCL Press, King's Lynn and Girl Ford Press, England, 1995.

Sinha, *Human Health and Environment*, Efficient Offset Printers, New Delhi, 1977.

Vyas S. Vijay, "Environmental Policies and Policy Implementation in India," in *Environmental Development* by Rathore, Rawat Publication, New Delhi, 1996.

11

Environment Management
A Study in Environmental Awareness

M.S. GILL and NAVDEEP KAUR

There is degeneration of the supportive and sustenance system due to excessive deforestation, building of hydro-electric projects, transportation and emission of greenhouse gases and to house the increasing population. The process of deforestation has surpassed all the previous growth rates and approximately 22 hectares of forest is cut down every minute on this earth (Pannersevam and Ramakrishnan, 1996). On the other hand, the survival rate of population has increased tremendously over the year with increase demand for foodgrains and life support machines while deforestation has been leading to increasing soil erosion, global warming and the like (Salgare and Iyer, 1993; Gupta, 2000) resulting in changed soil type and its productivity. The development path followed so far with the idea of growthmanship has aggravated the situation with increased emission of greenhouse gases into atmosphere, industrial effluents into the lithosphere and hydrosphere. According to a survey, 35 million people consume contaminated water; fish oil and meal have high level of dioxin and other toxins (Down to Earth, 2001, 2002). Thus, it is imperative to manage environment and for this, environmental awareness helps people and social groups to acquire sensitivity to the total environment and its related problems. Such awareness includes all the vital aspects like air, water, soil, and even noise pollution.

Further, every human society, be rural, industrial, urban or technologically advanced/advancing/underdeveloped, disposes certain by-products of degradable or non-degradable type. Biodegradable pollutants are manageable because of homeostatic qualities but there is a limit to which nature has capacity to correct the imbalances of the environment. The non-biodegradable pollutants are non-manageable as these continue to occupy landfills, beds of water bodies such as rivers, oceans, and the like. The presence of such pollutants often blocks the recharging of ground aquifers or lead to leaching of toxic contents into fresh water bodies rendering these unfit for human consumption. The excessive emission of such pollutants effects human life (Manivasakam, 1984; Saxena, 2000; Down to Earth, 2000). Moreover, discarded materials like TV, refrigerators, computers and many more are further adding radiation causing increased incidence of skin cancer. Such materials are usually dumped into the sea or exported to developing countries as grants without realizing their severe consequences. It is in this context, environmental awareness and its management becomes important. For this, environmental assessment has been made a part of each and every development project (Kulshrestha and Singh, 2000; Ghazala, 2001). Also if environmental awareness is a necessity of the hour, then it becomes pertinent to know the level of awareness among the growing educated class when studies by Aggarwal (2001), Venkateswaran (1995) and others indicate that urban educated class is not conscious of the environment or has a little concern.

Objective of the Study

This study attempts to know the level of awareness among the students at their formative stage of education. The beliefs and ideas incorporated at the school level shape not only the awareness about national and international issues, but also promote them to social action. For this, the main objectives of this study are:

(a) The awareness level towards important components of environment.
(b) The difference in the level of awareness as a product of their type of schooling.

Method of Study

Urban areas are considered as the centres of social, cultural and intellectual activities and as such efforts are made to provide avenues through which one can imbibe such behavioural patterns that help one to excel in these fields. The universe of the study is Amritsar city in Punjab. From the different types of schools, a Government and a private school were selected. A questionnaire comprising of 67 items was administered. Patil *et al.*, have standardized the questionnaire in their study conducted for students of Maharashtra. The data was collected from students of class IX present on the day of questionnaire administration. In all, 116 and 90 students from Government and private school filled up the questionnaire in the presence of the researcher. The researcher intervened to clarify the meaning of the terms wherever required without influencing the response of the students. The data so collected was computed and to calculate association, co-efficient of co-relation was used. Further, cumulative score for each respondent was calculated and compared.

Results and Discussion

Since the objective of the study was to find out the level of awareness among various components of environment, following components had been taken into consideration:

(a) Solid Waste

Solid waste has been defined as garbage from house, waste from industrial and commercial activities, waste from agriculture, mining and water purification plants and the like. According to Resource Conservation and Recovery Act of US, solid waste is "any garbage, refuse, sludge from water treatment plants, on air pollution control facility and other discarded material, including solid, liquids, semi-solid or contaminated gaseous material resulting from industrial, commercial, mining, agricultural operations and from community activities, but does not include solid or dissolved material in domestic sewage (cited from Anand, 2005). In fact, it includes all those materials that are discarded, unusable (may be reused), desirable or undesirable. It is estimated that an individual generates 0.3 to 0.5 kg. of solid waste per day. Urban areas are facing serious challenge with regard to waste generation and its disposal. Based on 500 gram of waste per capita generation, it is estimated that Delhi generates about 6500 tonnes of waste and it is disposed of in low-lying areas at the landfill sites (Kumar, Kumar and Parmar, 2005). It is the same story of other cities through out of India. In this way, landfills are packed with all sorts of non-bio-degradable, bio-degradable and other toxic wastes. Further, rules exclude bio-medical waste, solid waste, water waste and exhaust gases from hazardous waste as per the draft hazardous material rules of the Union Ministry of Environment and Forest (Kaur, 2007). Moreover, in the present scenario, it is the constitutional responsibility of the governing bodies to work for proper disposal of waste. Non-treatment of solid waste and its lying in open landfills act as breeding ground for pests, insects and rodents besides emitting foul smell more so during summer and rainy seasons creating climate problems. It contaminate water bodies through leaching as well. Here, the emphasis is on waste generated by the city/municipal waste. Also, the respondents were asked whether they generate more waste than what is stated above. Approximately 30 per cent of the respondents have stated that they generate more than the permissible one. Not only this, 85 per cent of the respondents have offered to work for creating awareness among the people for environmental protection. In all, 21 questions were asked. The total score varied from 0 to 21. The study reveals that majority of the respondents had score of 17; only 6.3 per cent of them have a score of 20 (Table 11.1). Further, 2.4 per cent of the respondents had a score of 11, which is the lowest score and all such respondents were from the private school. Another 1 per cent who got 12 score on this account too were from the private school. As the awareness score increased from 17 onwards, the representation of public school increased from that of private school respondents.

It reflects that the awareness level with regard to solid waste is high among the students studying in Government institutions as compared to those in private institutions. Moreover, the awareness score of most of the students from private school varied between 16 to18 and that of public one from 17 to 19.

Table 11.1: Awareness Score towards Solid Waste

Score	*Number of Respondents*		*Total*
	Public	*Private*	
11	0 (00.0)	5 (05.6)	5 (02.4)
12	0 (00.0)	2 (02.2)	2 (01.0)
13	1 (00.9)	6 (06.7)	7 (03.4)
14	2 (01.7)	6 (06.7)	8 (03.8)
15	8 (06.9)	9 (10.0)	17 (08.2)
16	16 (13.8)	19 (21.1)	35 (17.0)
17	26 (22.4)	18 (20.0)	44 (21.3)
18	28 (24.1)	13 (14.4)	41 (20.0)
19	26 (22.4)	8 (08.9)	34 (16.6)
20	9 (07.8)	4 (04.4)	13 (06.3)
Total	**116 (56.3)**	**90 (43.7)**	**206 (100.0)**

(b) Air Pollution

Air pollution refers to excessive concentration of foreign matter in the air which adversely affects the well-being of individual and other biotic life. Air pollution is a serious threat to human existence as it is responsible for a number of diseases besides change in climate. Pollutants in the air affect respiratory system the most followed by central nervous system. The two significant sources of it are growing number of industries and vehicles, which emit gases including toxic ones. According to Walther (2006), with increase in energy demand emission level would increase by more than half (52 per cent) from today's level. Three quarter of this (73 per cent) come from developing countries. This indicates the level of carbon dioxide, which is going to bring about global changes in environment, agricultural production and health problems. Anand (2005) finds thermal power plants in Delhi contribute 13 per cent and industrial activities about 12 per cent of total air pollution. The transport sector in India accounts for around 18 per cent of India's total carbon dioxide emissions (Sharma, 2007). Carbon dioxide emissions are bound to increase with expansion of auto-industry. Carbon dioxide is the first and major greenhouse gas responsible for global warming followed by methane (Tandon *et al.*, 2005). The burning of paddy fields after harvest is the cause of serious bronchial and other ailments. It is important for the corporate sector and individuals to consider their social responsibility to check emission level responsible for global warming and health related problems. Keeping this into consideration, twelve items were administered on this topic. The cumulative score of 12 indicates high level of awareness. High level of awareness was found more among the private school respondents than that among the public school respondents (Table 11.2). Median score on this component was 9 but majority of the students had an average score of 10. Those securing awareness score of 10 were to a large extent equally represented from the public and the private school. As awareness of air pollution increased from 11 onwards, the proportion of students from private school in comparison to public school respondents increased. The proportion of students having less than 8 score had been found to be high among the students belonging to the Government school, thereby suggesting that awareness level towards air pollution among the students of private institution is more than that among the Government school. Not only this, most of the students from public school had awareness score ranging from 6 to 8 whereas that of the private school had 10 to 12. This may be attributed to their confrontation in day-to-day life with such problem as this school is located on the main transport route where traffic flow is usually high.

Table 11.2: Awareness Score towards Air Pollution

Score	*Number of Respondents*		Total
	Public	*Private*	
5	1 (00.9)	0 (00.0)	1 (00.5)
6	18 (15.5)	2 (02.2)	20 (09.7)
7	20 (17.2)	3 (03.4)	23 (11.1)
8	23 (19.8)	10 (11.1)	33 (16.0)
9	21 (18.1)	16 (17.8)	37 (18.0)
10	21 (18.1)	20 (22.2)	41 (20.0)
11	10 (08.6)	29 (32.2)	39 (18.9)
12	2 (01.7)	10 (11.1)	12 (05.8)
Total	**116 (56.3)**	**90 (43.7)**	**206 (100.0)**

Noise Pollution

Noise pollution has many sources like pressure horns, automobiles, industries, loud speakers, playing of audio system on very high pitch and the like. The acceptable level of noise for human health ranges from 40 to 140 dB depending upon the source. Students found that pressure horns, loud speakers and high pitched audio systems are the causes of noise pollution and most adverse effect of it is hearing impaired. This has psychological and physiological consequences by disturbing peaceful atmosphere and sleep: more so among the old, infirm, infants and sick persons. Vehicles particularly, auto-rickshaws have been identified as major source of noise pollution. Anand (2005) finds that in addition to the above, Diesel generating sets as major source on noise pollution in Delhi. In all, 8 items were administered on this issue. Nearly half of the respondents had got the highest score that is 8 (Table 11.3). Only 0.5 per cent of the total respondents had a score of 4 (the minimum one on this issue) and all of them were the students of public school. Further, 1.5 per cent of the respondents had awareness score of 5 and all were the students of private educational institute. In all, the proportion of respondents getting higher score on this issue was more among the public school than that of the private school. Also, the awareness score of the students from private school varied from 6 to 7 and that of the public school from 7 to 8, not a significant difference from the data point of view.

Table 11.3: Awareness Score towards Noise Pollution

Score	*Number of Respondents*		*Total*
	Public	*Private*	
4	1 (00.9)	0 (00.0)	1 (00.5)
5	0 (00.0)	3 (03.3)	3 (01.5)
6	6 (05.2)	10 (11.1)	16 (07.8)
7	55 (47.4)	32 (35.6)	87 (42.2)
8	54 (46.5)	45 (50.0)	99 (48.0)
Total	**116 (56.3)**	**90 (43.7)**	**206 (100.0)**

(c) Water Pollution

Water is an elixir of life, as life cannot sustain itself without water. People are generally unaware of the depleting resource of water and continue to use recklessly especially those in industrial and agricultural sector. According to OPEC Bulletin (2007), by the middle of this century at least two billion people—may be seven billion—could be short of this precious source. According to United Nations, 1.1 billion people are currently without safe drinking water, while 2.6 billion lack access to basic sanitation. Further, water borne diseases continue to cause deaths of thousands of children across the world, everyday. For some, the crisis means having to walk long distance everyday to fetch enough drinking water—clean or unclean. Not only this, water use rate has increased nearly six-fold since the beginning of 20th century in spite of the fact that only 10 per cent of the water is used for residential purposes. Major portion of the water in domestic use is of toilet. Agricultural sector consume roughly 70 per cent of the water for irrigation to produce sufficient to meet the needs of the growing population. With stress on raising productivity on farm more so among the developing countries whose economic base is crop husbandry, the use of water far exceed causing water scarcity. Shortage of water has given rise to another class of capitalist who sell it for capital generation. Navroj (2002) finds that shortage of water has given rise to private owners who capitalize on selling it to cultivators in Gujarat and Madhya Pradesh. In Punjab, due to excessive use of ground water, 103 blocks out of 138 blocks are over-exploited and 93 of them are declared 'dark' and another 13 'grey' (Virk, 2008). Likewise, people in metropolitans' trade in water especially during summer when there is acute shortage even of the drinking water. In India like many Third World countries, water is mostly contaminated and at some places even unfit for animal consumption. The reasons are sewerage and industrial effluents, which freely flow into streams and other water bodies. In Kolkata, it usually mix with drinking water pipes during rainy season rendering water unfit for consumption and spread of water borne diseases. In Abidjan, urban sewage is the main contributor contaminating ground water (OPEC, 2007). The untreated sewage and sewerage flowing in various open drains are one of the causes of ground water quality deterioration in Delhi (Kumar *et al.*, 2005). The survey of groundwater quality has indicated higher concentration of total dissolved salt, electrical conductivity, chlorides, nitrates and fluorides. Certainly, the respondents were aware of water contamination, effluents flowing in water bodies and its seepage to ground aquifers but were unaware of agricultural and other sources of water contamination. Twelve items were administered to know about the causes of water pollution and its effects. Most of the respondents held that consumption of contaminated water leads to a number of diseases especially during summer. They were aware of the use of chlorine for water purification. Nearly 7 per cent of the student respondents had the maximum awareness score of 12 (Table 11.4). Majority of them were having awareness score of 10 or 11. Only 0.5 per cent had awareness score of 4.The respondents of the private school had a lower score on water pollution as compared to that of the public school. The proportion of students having a score of 11 or 12 was more from the public institute as against those of the private one. The average score of most of the respondents from the private school varied from 7 to 10 whereas it varied from 8 to 11 for the students of public school.

Table 11.4: Awareness Score towards Water Pollution

Score	*Number of Respondents*		*Total*
	Public	*Private*	
4	0 (00.0)	1 (01.1)	1 (00.5)
5	0 (00.0)	0 (00.0)	0 (00.0)
6	0 (00.0)	3 (03.3)	3 (01.5)
7	1 (00.9)	4 (04.4)	5 (02.5)
8	10 (08.6)	10 (11.1)	20 (09.7)
9	22 (19.0)	25 (27.8)	47 (22.8)
10	33 (28.4)	25 (27.8)	58 (28.1)
11	44 (37.9)	14 (15.6)	58 (28.1)
12	6 (05.2)	8 (08.9)	14 (06.8)
Total	**116 (56.3)**	**90 (43.7)**	**206 (100.0)**

Conclusion

The capitalist and socialist growth models irrespective of their doctrine have raised gross national product and per capita income with the exploitation of natural resources. The technical advancements have laid the road to progress but not without a price in terms of depletion of resources and erosion of protective and sustenance systems. A debate was initiated to transform the negative urges of development in 1972 at the national and international level with the active involvement of formal and informal institutions. Their efforts are directed towards creating environmental awareness and programme to realize the socio-economic effects such degradation would lead to. How far such initiatives have made their impact in the conservation and awareness among the budding generation is the main theme of this paper. Though there is not much variation, the study infers that coming generation is getting abreast with the problem of environment and its effects. The awareness score of the Government and private school respondents with regard to air pollution varies much than that on other components. The awareness level of the private school students is much higher on the air pollution component than that of the government school. The study highlights that the students of the government school are much aware of the problem of solid waste, noise pollution and water pollution than the students of private school. Still, the students are catching up with the issue and are becoming aware of it. Mehta (2004) has corroborated the role of youth in generating environmental awareness and their increasing consciousness about environmental degradation. Nanda (2005) too has stressed the need of mass societal participation in environment awareness and environment literacy and the present study points out the growing environment awareness among the youth studying in schools. The variation is more due to one's immediate interaction with such components. The study by Heart Care Foundation corroborates our findings. The variation can be attributed to the policy of government to introduce and conduct environment campaigns and environmental programmes in the government institutes. Marten (2005) holds that where top-down regulations and high priced technical fixes are not working, positive environment tipping points offer a third way to restore communities, both natural and human. So, environmental awareness at local level within the community may rejuvenate environmental issue reverting negative trends. Kumar *et al.* (2005) too corroborates this finding in the context of quality of water. Educational institutes therefore play significant role in realizing goals of creating environmental friendly society. Of course, the role of mass-media cannot be underscored.

REFERENCES

Aggarwal A. "Common Property, Forest Management and Indian Himalayas", *Contribution to Indian Sociology*, Vol. 35, No. 2, 2001.

Aggarwal Manmohan, *Globalization and Millennium Development,* Social Science Publication, New Delhi, 2007.

Aggarwal and Dubey, *Environmental Controversies,* APH Publishing Company, New Delhi, 2002.

Anand S. "Solid Waste Generation and Management in Delhi: A Sustainable Approach", in *Environment and Development* by Jagbir Singh (ed.), I.K. International Pvt. Ltd., New Delhi, 2005.

Chahal S.K., *Environment and World Life: Towards a New Paradigm,* Ashish Publishing House, New Delhi, 1994.

Economic Survey, a Ministry of Finance Publication, Government of India, 2004.

Ghazala S. "Forest Fragmentation, that what is causing Bio-diversity Losses", *Down to Earth*, Vol. 9, No. 16, 2001.

Ghosh A., *Urban Environmental Management: Local Government and Community Action,* Concept Publishing Company, New Delhi, 2003.

Gupta S. *Global Environment*, Sarup and Sarup Publications, New Delhi, 2000.

Human Development Report 2006, United Nations Development Program Publication, UN Plaza, New York, 2006.

Jaswal, P.S., *Environmental Laws,* Pioneer Publications, New Delhi, 2006.

Kaur, R. "Waste Material", *Down to Earth*, Vol. 16, No. 15, 2007.

Kulshreshtha and G. Singh, "System of National Account and Accountability", in *Environment Management and Audit* by Rao and Rao, Deep and Deep Publications, New Delhi, 2000.

Kumar, H.D., *Sustainable Human Ecology,* Atlantic Publishers, New Delhi, 2002.

Kumar Praveen *et al.* "Ground Water Quality and its Pollution: A Case Study of Delhi", in *Environment and Development* by Jagbir Singh (ed.), I.K. International Pvt. Ltd., New Delhi, 2005.

Kumar, Kumar and Parmar "Solid Waste Management: A Contemporary Issue for Urban Areas", in *Environment and Development* by Jagbir Singh (ed.), I.K. International Pvt. Ltd., New Delhi, 2005.

Lee A. James, *Environment, Public Health and Human Ecology,* John Hopkins University Press, London, 1985.

Manivasakam N. *Environmental Pollution*, National Book Trust, New Delhi, 1984.

Marten, Gerald *et al.* "Environmental Tipping Points: A New Slant on Strategic Environmentalism", *World Watch Magazine*, 2005.

Mehta R. "Views of the University Youth on Environmental Sustainability", *Research Journal Social Sciences*, Vol. 12, No. 223, 2004.

Nanda K. "Environment Awareness, Education—Some Suggestions", in *Environment and Development* by Jagbir Singh (ed.), I. K. International Pvt. Ltd., New Delhi, 2005.

Navroj D.K. *Tubewell Capitalism: Ground Water Development and Agrarian Change in Gujarat*, Oxford University Press, New Delhi, 2002.

Pannerselvam A. and M. Ramakrishnan *Environmental Science Education*, Sterling Publishers Pvt. Ltd., 1996.

Robertson R., *Globalization: Social Theory and Global Culture*, Sage Publishers, London, 1992.

Salgare S.A. and M.P. Iyer, "Environmental Chaos" in *Environmental Ruin* by R.M. Lodha, Indus Publishing Company, New Delhi, 1993.

Saxena, H.M. *Environmental Management*, Rawat Publications, New Delhi, 2000.

Sharma, Vibhu-Climate Change put on top of World Agenda, *The Tribune*, Dec. 27, 2007.

Speth, L.J., *Global Environmental Challenges: Frontiers to a Sustainable World*, Orient Longman, New Delhi, 2006.

Tandon, M. *et al.*, "Methane Production, Global Warming and Ozone Layer Depletion" in *Environment and Development* by Jagbir Singh (ed.), I.K. International Pvt. Ltd., New Delhi, 2005.

Venkateshwaran S. *Environment, Development and Gender Gap*, Sage Publications, New Delhi, 1995.

Virk, B.S. Water and Power Crisis in Punjab, *The Tribune*, Jan. 5, 2008.

Walther A. "Horizons of Dialogue", *OPEC Bulletin*, Vol. xxxvii, No. 2, 2006.

———"Gas Chambers", *Down to Earth*, Vol. 10, No. 2, 2001.

———"Environmental Science", *Down to Earth*, Vol. 10, No. 19, 2002.

———"Water Resource Management: Making every Drop Count", *OPEC Bulletin*, Vol. xxxviii, No. 2, 2007.

———"IAEA Symposium hears how Hydrology can enhance carbon capture", *OPEC Bulletin*, Vol. xxxviii, No. 5, 2007.

12

Environmental Law and the Indian Judiciary

A Mechanism of Facilitations and Inhibitions (A Socio-Legal Analysis)

JASLEEN KEWLANI and NISHANT SHARMA

Introduction

Environmental degradation is a global issue because 'environment' is such a context which cannot be restricted to geographical boundaries. Amidst globalization of the issue of environmental degradation, its acute state in India cannot be ignored. India's environmental problems are gaining global significance because of the rapid and offensive speed of urbanization and lack of infrastructure. Increasing urbanization, industrialization, deforestation and transportation as the other cousins of economic development put tremendous pressure on natural resources, and therefore, there is a pressing need to strike a balance between developmental planning and urgency to safeguard the environment (Ghosh, 2007). As asserted by through an environmental analysis, there are ten drivers of Environmental Deterioration. Those are: (1) population, (2) affluence, (3) technology, (4) poverty, (5) market failure, (6) policy and political failure, (7) the scale and rate of economic growth, (8) the nature of our economic system, (9) our culture and its values, and (10) the forces loosed upon the world by the globalization of the economy (Speth, 2004). All these forces lead to misuse and over-exploitation of the natural resources, which is a danger to preservation of the environment.

'Globalization' as a concept refers to the compression of the world and the intensification of the consciousness of the world as a whole (Robertson, 1992). When consciousness of the world gets intensified, it gives bounce to the quest for development; which encapsulates multiple angles of survival including that of the environment also. The process of development under globalization implies hyper-consumption of energy and natural resources, leading to environmental degradation. This kind of imbalanced development further results into emergence of many social inequalities and issues. In such an adverse situation, it is only ten per cent of the world's population which is consuming the bulk of non-replenishable natural resources for satiating the avarice of the developed world at the cost of ninety per cent of the world population. This reflects a threatening phenomenon in which the present generation of the world is not only damaging nature and its environment against the interests of contemporary relatively less developed populations but also against the interests of their own future generation (Aggarwal, 2007). Thinking the situation to be a grave one, it is crucial to give an adequate attention to the environmental concerns; devoid of any bias for the motive of economic development. In a complementing perspective, the paper forwards some relevant contents. The contents include: (i) The Role of Judiciary and the Environmental Law (with a specific emphasis on Annual Survey of the year 2006); (ii) Administrative and Legislative Measures for Environmental Protection; and (iii) The Non-Judicial Responsibility and Environmental Protection.

The Role of Judiciary and the Environmental Law

At the very outset, the preamble of the Constitution of India states that the functioning and governance of the nation will be carried ahead on 'Socialistic' pattern which impliedly fosters equilibrium in distribution of natural resource also. Moreover the spirit of the word 'Socialistic' entails that social problem will be given a higher consideration as compared to that of the individual problems. Other crucial content of the preamble is reflected in the term 'Democratic' which provides noticeable power to the common masses to act with sovereign attitude and provides them a right to participate in Government's decision-making process also. In apposition to

Democracy, 'Federalism' has also been adopted, which provides independent powers to local governments to maintain the environment. Article 21 Part IV of Constitution of India consisting of Fundamental Rights, guarantees a fundamental right to life; a life of dignity and a life which is to be lived in a proper environment, free of danger of disease and infection. No justification is needed here because no one can lead a healthy life without a healthy and hygienic environment. Not only this, it has been asserted that 'right to live in healthy environment' is the *'sanctum sanctorum'* of Human Rights also (Jaswal, 2006). This abstraction was given a manifest implication in a case named **"R.L. and E. Kendra, Dehradun** v. **State of U.P. (AIR, 1985).** In this case, the right to live in healthy environment was given an imminent importance. The plaintiff pleaded before the Supreme Court against the progressive mining, which denuded the hills and natural capital of Mussoorie and attacked trees and forests of this region, which further led to uncontrolled soil erosion. The court treated this as a writ petition under **Article 32** of the Constitution (epistolary jurisdiction). Hence the right to protect and preserve the environment was held high.

Another exemplary litigation in this context is **T. Damodar Rao** v. **S.O. Municipal Corporation, Hyderabad (AIR, 1987).** While providing decision in this case, the court pointed out that in view of Article 48-A and 51-A(g), it is made evident that protection of environment is not only the duty of citizen but it is an obligation of the State and all enforcement agencies of the State too. This adjudication forwarded environmental directives to States also.

A complementary attempt can be noticed in **Kinkri Devi** v. **State (AIR, 1988),** in which the court pointed out that it is both the Constitutional Directive to the State and a Constitutional Duty of the citizens too that they will not only protect but maintain and improve the environmental quality along with natural flora and fauna like forests, rivers, lakes etc. This decision laid another milestone to encourage the common masses to initiate mass improvements on individual level.

In **M.C. Mehta** v. **Union of India (AIR, 1987).** which is also commonly known as Oleum Gas Leakage Case, the Supreme Court forwarded the decision providing an inferred implication that the right to live in a pollution free environment is a part of Fundamental Right to life under Article 21 of the Constitution. The reason for this judgement can be termed as, constitutionally driven because slow poisoning by the polluted atmosphere caused by environmental pollution and spoliation should also be regarded as 'amounting to violation of Article 21 of Constitution of India', as it endangers life of the suffering individuals.

One among such protective measures rendered for environmental maintenance, is **L.K. Koolwal** v. **State (AIR, 1988).** The adjudication made in this case is in synonymity with that of the above discussed case. Its decision forwarded an assertion that maintenance of health, preservation of the sanitation and environmental protection also falls within the purview of Article 21, because a worse environment as it adversely effects the life of citizens and gradually poisons their existence.

In **Subhash Kumar** v. **State of Bihar (SCC, 1991),** the Supreme Court added that the right to the enjoyment of pollution free water and air is a credential for the full enjoyment of life, which is provided under Article 21 of Constitution of India. It provided all citizens a direction to have resort to Article 32, if anything endangering occurs to the environment around them.

While making a slight move back to the above discussed cases, **Charan Lal Sahu** v. **Union of India** is also a landmark in the history of environmental protection. Reference of Article 21, 48-A and 51(A)(g) was made in context of Human Rights in this case. Here it was stated that right to have a clean and healthy environment is one of the Human Rights of all the citizens. This notion was upheld while showering an indepth consideration for the validity of Bhopal Gas Leak Disaster Act, 1985.

Some other examples of such cases filed to protect and claim one's environmental rights are; **"Indian Council for Enviro-Legal Action** v. **Union of India,**[1]"; **"Vellore Citizen's, Welfare Forum** v. **Union of India,**[2]"; **"Dr. Ashok** v. **Union of India**[3]".

In the coming part of the paper, Annual Surveys of the years 2006 and 2007 have been incorporated. These Annual Surveys deal with the Environmental Law. This part of the paper reflects the role of judiciary in strengthening the environmental conscience of the mankind by the way of judicial potential. The cases dealt here or incorporated hereby show the process of determining 'priorities'. The

Judiciary of the nation put efforts to balance the environmental and economic interests of the masses. With the help of the Annual Surveys given below, an attempt has been made to analyse the role of Judiciary in managing and preserving environment.

Annual Survey, 2006

In an attempt to establish the principle of 'Sustainable Development', **Mrs. Susetha** v. **State of Tamil Nadu**[4], is a landmark judgement. This judgement makes the principles and components of Sustainable Development clear and properly implied. The appellant herein claims to be member of the Okkiam Thoraipakkam Panchayat Union. There exists a temple tank in the small village. The said tank admittedly was lying in disuse and was an abandoned one. The Panchayat took decision of constructing a shopping complex for the purpose of user thereof for resettlement of there persons who were displaced due to the expansion of high way project launched in the same area. The State of Tamil Nadu also issued a Government order permitting the construction of a shopping complex therein. The writ petition filed by the appellant was first dismissed by Madras High Court and then moved to Supreme Court. The judgment of the Supreme Court includes that water bodies are required to be retained. Such requirement is envisaged not only in view of the fact that the right to water as also quality life are envisaged under Article 21 of the Constitution of India, but also in view of the fact that the same has been recognized in Articles 47 and 48-A of the Constitution of India. Article 51-A of Indian Constitution furthermore makes a fundamental duty of every citizen to protect and improve the natural environment including forests, lakes, rivers and wildlife. But as per the Supreme Court's order, this principle applies only to natural water storage resources. The same can't be applied in relation to artificial tanks. Where a temple tank which has lost its utility long back and now used as dumping yard, was used by Panchayat for constructing a shopping complex for the purpose of user thereof for resettlement of those persons who were displaced due to expansion of highway project. More so, when the village in question is situated near a sea and having fine water tanks in or around therein. It is clear from this judgement that threat to sustainable development is realized if an existent natural unit, which is in utility, is endangered. If an abandoned natural body or unit is being utilized for the purpose of economic development, it does not amount to threat to sustainability. Recently in **T.N. Godavaraman Thirumulpad** v. **Union of India**[5] supported the fact. This judgement highlighted the importance of preservation of natural lakes and in particular those which are protected under the Wild Life (Protection) Act, 1972. In **Intellectual Forum Tirupate** v. **State of A.P. and Others**[6], it was held that it is important to protect and improve the natural environment including forests, lake, rivers and wildlife. This case dealt with the State's duties with regard to public trust. If a State holds a resource that is freely available for the use of the public, it provides for a high degree of judicial scrutiny on any action of the Government, no matter how consistent with the existing legislations, that attempts to restrict such free use. To properly scrutinize such action of the Government, the courts must make a distinction between Government's general obligation to act for the public benefit, and the special, more demanding obligation, which it may have as a trustee of certain public resources. In **Bombay Dying and Mfg. Co. Ltd.** v. **Bombay Environmental Action Group and Others**[7], reference was made to a large number of decisions and it was said that whereas the need to protect environment is a priority, it is also necessary to promote development stating; the harmonization of two needs has led to the concept of sustainable development, so much so that it has become so significant and focal point of environmental legislations and judicial decision relating to the same. Sustainable Development, simply put as a process, in which development can be sustained over generations. It was also added that the concept of sustainable development operational for public policies raises important challenges that involve complex synergies and trade offs.

Another case which added to the credibility of the concept of 'Sustainable Development' is **Karnataka Industrial Areas Development Board** v. **C. Kenchappa and Ors.**[8] The importance and awareness of environment and ecology is becoming so vital and important that court makes it imperative on State Industrial Areas Development Board to insist on the condition emanating from the principle of Sustainable Development. In this case the court issues following directions; that in future before acquisition of land for development, the consequence and the adverse impacts of development on environment must be properly comprehended and lands be acquired for development that they do not gravely impair the ecology and environment; State Industrial Area Development Board to incorporate the condition of allotment to obtain clearance from the Karnataka State Pollution Control Board before the land is allotted for development. The said directory condition of allotment of lands be converted into a mandatory condition for all the projects to be sanctioned in future. This judgement also made a detailed reference to Section 3 of Environment Protection Act (1986).

Water and Environmental Management

At the start of the 21st century water borne infectious diseases are thing of the past in rich countries, accounting for a fraction of 1 per cent of overall mortality. At the turn of the 19th century, diseases like diarrhea, dysentry and typhoid and fever posed major threats (Human Development Report, 2006). In late 19th century, they accounted for 1 in 10 deaths in US cities, with children the primary victims. Not only this rather many water related issues have been a talk of globalization because all the nations suffer from same or other forms of water related crises.

Scarcity of water, pollutants in water, water-sanitation, are such issues which seek attention of the common masses along with the governance. In similar context many judicial proceedings had been undertaken. In intellectual forum, **Tirupathi** v. **State of A.P. and Others.**[9] Two kinds of questions were raised. Firstly, at a jurisprudential level, it falls on this court to lay down the law regarding the use of public lands or natural resources, which have a direct link to the environment of a particular area, by the Government. Secondly, this court should decide, on the facts of the present case, the order to be passed with respect to two tanks. The present case related to the preservation and restoration of *status quo ante* of two tanks, historical in nature being in existence since the time of Srikrishan Devarya, 1500 A.D. the names of the tanks are 'Avilala' and 'Peruru.' The grievance of the appellant was the systematic destruction of percolation, irrigation and drinking water tanks in Tirupathi Town, namely 'Avilala' and 'Peruru tank' and alienation of the Avilala tank bed land to Tirupathi Urban Development Authority (TUDA), and A.P. Housing Board and Peruru Tank bed land to Tirumala Tirupathi Devasthanam (TTD) for housing purpose. According to appellant, the cry of socially spirited citizens calling for judicial remedy was not considered in the right perspective by the Division Bench of High Court of Andhra Pradesh.

Despite there being overwhelming evidence of the tanks being in existence and were being put to use not only for irrigation purpose but also as lakes which were furthering percolation to improve the ground water table, thus serving the needs of the people in and around the tanks; it was submitted that the high court has given precedence to the economic growth by completely ignoring the importance and primacy attached to the protection of environment and protection of valuable and most cherished fresh water resources. In this case environment side won the battle. Supreme Court afterwards referred to Constitution of India, Arts. 21, 48A, 51A, Environment Protection Act (29 of 1986). According to the Supreme Court, it is important to maintain a balance between environmental protection and social development. Tanks are important part of environment of an area and government is responsible to protect and preserve historical tanks *qua* concept of 'Sustainable Development' and 'Public Trust Doctrine'. Merely asserting an intention for development is not enough to sanction destruction of ecological resources. Property subject to trust must not only be used for a public purpose but it must be held available for use by general public. Property must be maintained for particular type of use (i) either traditional uses, or (ii) some uses particular to that form of resources—Principle of 'Inter-generational Equity' also to be applied for protecting natural resources. Constitution of India, Article 21—Right to Shelter *qua* preservation of historical tanks/environment—Tank is an environmental property and State authorities are trustees to hold and manage such properties for benefits of community—State cannot be allowed to commit any act or omission which will infringe right of community and alienate property to any other person or body. No one will be left harmless if proposed construction are not carried out and Right to Shelter not so pressing so as to outweigh all environmental considerations. The fact that the party has spent money on developing land is immaterial. Court passed directions with regard to feasibility of revival of 'Peruru Tank' and 'Avilala Tank' in Tirupathi and prohibited further housing construction.

In **Mullaperiyar Environmental Protection Forum** v. **Union of India and Others**[10], another water related issue was decided. States

Reorganization Act (37 of 1956), S. 108—Constitution of India, Arts. 3, 4; Sch. 7, List 2, Entry 17 deals with the Right to receive and utilize water. Inter-State agreement between the predecessor State relating to irrigation and power generator etc. continues, but State cannot claim to have legislative powers over such waters. Hence, Section 108 stands valid. Constitution of India, Act. 363—deals with Inter State dispute relating to safety of dam—Arising out of agreement for use of water between Maharaja of Travancore and State Secretary for India. Dispute is not in respect of right accruing or liability or obligation arising under any of provisions of Constitution, and jurisdiction of Supreme Court to adjudicate each dispute not ousted *qua* Art. 363. Constitution of India, Art. 32 Arbitration and Conciliation Act (26 of 1996), Section 7 deals with alternate remedy of arbitration. It was stated that controversy related to increase in water level in reservoir is not about rights, duties and obligations or interpretation of any part of inter-state agreement relating to use of water. Hence, dispute not liable to be referred to arbitration and discretionary relief can be granted under Art. 32. Another reference was made in this case. It was Forest (Conservation) Act (69 of 1980), Section 2—Wildlife (Protection) Act (53 of 1972), S. 26A—deal with environment and safety of people. Raising of water level of Mullaperiyar dam from 136 ft to 142 ft will not alter the boundaries of sanctuary and will not effect wildlife or forests or flora and fauna. Strengthening work of existing dam in forest is not non-forestry activity, and due to this, environment is not adversely affected. That apart, there is no report to suggest of dam being jeopardized on raising of water level—hence, water level is permitted to be raised to 142 ft.

Air Pollution

Environmental degradation has resulted into an immeasurable addition of pollutants in air, which further produces multiple adversities for physical well-being of individuals. Transportation is an integral part of increasing population. Where human settlements and housing increase, the need of accessibility to these establishments also gain momentum. Transportation is meant to have an access to the establishments. In developed and also in developing nations like India, many people prefer having personalized transport, but there is an administrative shift towards public transport. Public transport is preferred, as this is the only way to effectively deal with energy crisis, congestion and other environmental concerns. It can carry a large number of people over longer distances using lesser road space and fuel per person and at the same time causes less pollution (Kumar, 2002). Hydrocarbon emissions from these vehicles are quite high and about 20-25 per cent of the fuel used is lost this way. It creates air pollution and air pollution is associated with higher incidence of respiratory ailment, such as, bronchitis and asthma. This is the beginning; the influences get aggravated in later stages. The problems related with air pollution got exaggerated so much that they forward threat for physical well-being of individuals and their glib survival too. In context of air pollution sundry litigations have been a landmark in the year 2006.

In **Hetalben Jitendra Kumar Vyas** v. **Police Inspector, Sabarmati Police Station, Ahmedabad and Others,**[11] Environment (Protection) Act (29 of 1986), Section 25 and Environment (Protection) Rules (1989), Rule 89, were referred to. The contents deal with banning or regulation of a burning of crackers on public streets. There are sufficient legal provisions regulating manufacturing and bursting of fire crackers and therefore, any policy is not necessary. However, considering fact that the said provisions are not known to people at large and therefore, the said provisions are often violated, Secretary home department, directed that he should give direction to the concerned officers, who are concerned with the procedures of providing licences for manufacturing or dealing in fire crackers that fire crackers manufactured or sold by them should not be such, which would make noise above level prescribed in aforesaid rules. In this case, a writ petition was filed under Art. 226 of Constitution of India in context of payment against damages in care of bursting of crackers by members of a marriage procession on Public Street. One of fire crackers hitting right eye of the petitioner's son aged two and a half years, resulting in permanent loss of eyesight of his right eye. Claim of damages, i.e. a sum of Rs. 15,000 was awarded as costs/damages to petitioner payable by parents of bride and bridegroom jointly and severely. It was said that sum not awarded towards total damages and the claimant was entitled to initiate appropriate legal proceedings and recover damages. Though in this case, firing of crackers has led to a direct physical damage to a child; otherwise also, it is that human activity which pollutes environment and results into emergence of multiple respiratory and physical ailments.

In **M/s Alloy Steel Rolling Mills** v. **West Bengal Pollution Control Board and Others**[12], issue related to air pollution was discussed. Air (Prevention and Control of Pollution) Act (14 and 1981), Section 19, 53—Environment Pollution Rules (1986) Rule 3(2) were referred to. The complaint was about pollution by Alloy Steel Rolling Mill. Directions were given by Board that mill should operate only by using cleaner fuel, namely oil or gas instead of coal fired system. The directions were given by the Board after recording reasons and these directions were based on report of the Expert Committee and recommendation of national body which was backed by the suggestions of National Environment Engineering and Research Institute (NEERI) and in furtherance of public interest, hence, not liable to be interfered with, more so, in absence of proof of any bias or perversity in such directions.

Forests—A Crucial Ecological Unit

Environment can be managed if all the entities of environment are managed simultaneously, which refers to a planned and systematic approach for protection of land, air, water, forests, etc. All the environmental entities hold a vital position, but forests are the most productive and protective unit of ecology. Change in forest cover and its structure certainly affects air, water and other allied elements that is why forests are considered an important concern of environmental protection. Forests act as 'Carbon sink' which absorbs a large quantity of carbon and produces oxygen for the sake of humanity. If deforestation takes place it makes air easily prone to pollution leading to disastrous effects on mankind (Aggarwal, 2002). Many cases have been decided in context of either protecting forests from being exploited or in context of controlling the unregulated misuse of forest products. One of such judgements is **Ashok Kumar Anandani and Others** v. **The State of U.P. and Others.**[13] It falls under contextual premises of Forest Act (16 of 1927), Section 2(4)(b) (IV) and U.P. (Transport of Timber and other Forest Produce) Rules (1978) Rule 2—deals with levying transit fee on forest produce. The validity was established by stating that goods which are brought from forests and which cross forest during transportation would be covered under the definition of 'Forest Produce' under Section 2(4)(b)(IV). Thus transit fee is payable on stone ballast, Khanda and Boulder if they cross forest during transportation. Petitioners were using National Highway for transporting goods and there were several forests block on that highway. The court ordered that levy and demand of transit fee would be perfectly within the four corners of Transit Rules. The petitioners claimed that they are lifting and transporting the mineral accompanied by Form MM 11 after depositing the royalty and other expenses in accordance with the provisions of the rules and therefore, the respondent authorities cannot charge any transit fee on the same in pursuance of the provisions of the U.P. (Transport of Timber and other Forest Produce) Rules, 1978 (herein referred to as the Transit Rules). But the court said that the goods pass through forest in the course of transit hence the transit fee is to be levied and the petition was declared dismissed. This judicial proceeding shows that courts have a consideration for environmental related issues and the rules related to it are thereby adhered to very strictly.

Another case related to environmental concern is **Naresh Kumar Jain** v. **Deputy Commissioner, Hazaribagh and Others**.[14] In this case, the petitioner had prayed for a declaration that he is not required to take prior approval of the Central Government under the Forest Conservation Act, 1980 (herein after referred to as 'the Act') for his lease hold area. He had further prayed for quashing the office order contained in Memo No. 1751 dated 26-10-1996 (Annexure 7) whereby his application for grant of second renewal of the stone mining lease with respect to a portion of the leasehold are i.e. Plot No. 711 of an area of 4.50 Acres of Hazaribagh, Thana No. 250 at village Bamhani, District Hazaribagh was rejected. With reference to the Forest (Conservation) Act (69 of 1980) Section 2, the renewal of the mining lease was rejected on the ground that area in question was forest area and permission of Central Government as required under Act not obtained. It was added that petitioner had already applied for permission through forest department which was not yet rejected. In circumstances, petitioner was entitled to continue mining with permissible area till communication of order passed on his application.

In **M. Prabhakar Reddy** v. **State of Andhra Pradesh and Others**[15], the appellant (herein called, the plaintiff) filed a suit being O.S. No. 72 of 1988 on the file of the court of the 2nd Additional Subordinate Judge, Warangal, for recovery of an amount of Rs. 20,295 with interest by declaring that the search, seizure and collection of compounding fee of Rs. 20,000 is illegal. In his paint, he alleged that he purchased teak wood in two lots, got it sawn into sizes at two different saw mills and as he could not get permit, kept the cut teak wood at Ladella village with his cousin till 10-2-1988. When he was transporting the teak, the forest officials stopped the bullock carts, registered a case being P.O.R. No. 79/13/87 W/N dated 20-2-1988, and imposed compounding fee. He alleged that seizure and collection of compounding fee is illegal and therefore, he is entitled to such declaration and refund. The court referred to Andhra Pradesh Forest Act (1 of 1967) Section 2(g)—and Andhra Pradesh Forest Produce Transit Rules (1970), Rules 3, 4—It stated that there is a requirement of 'Permit' for transport of saw timber within the State. Rule 3 prohibits movement of 'Forest Produce' within the State unless accompanied by permit. It cannot be said that definition of 'Forest Produce' under Section 2(g) and Rule 3 excludes sawn timber. Finding of the trial court stated that no transit permit required in case of transit of sawn teak wood is illegal, hence, the appeal was dismissed. This proceeding shows the pivotal judicial concern for forest produce and their regulated use and transportation.

Another landmark judgement is **T.N. Godavarman Thirumulpad** v. **Union of India and Other**.[16] This case referred to the applicability of Public Interest Litigation. Here the court referred to Constitution of India, Arts. 226, 32, under which PIL can be filed in a case of general public welfare in High Courts and Supreme Court respectively. The court stated that the *Bonafides* of litigant petitioners are doubtful. As yet, in exceptional cases court may examine issues having regard to serious nature of public cause by appointing *'Amicus Curiae*—Central Empowered Committee (CEC) and *Amicus Curiae* appointed to assist court in determining issues relating to depletion of forests and preservation and conservation of forests. These kind of special committees demand due conformity and honour. Hence, the court stated that the approach used by the petitioner, in making insinuations against committee and *Amicus Curiae* was deprecated. The court, in the year 2006, 'not only itself toiled for environmental protection and preservation rather it also empowered the relatively appointed committees for the same purpose'.

Cases on Environmental Law (2007)

Research Foundation for Science, Technology and Natural Resource Policy v. **Union of India (UOI) and Ors.**[17] Alang is located on the west coast of Gujarat. It is the largest ship recycling yard in the world. It is one of the choicest ship-scrapping destinations for the ship owners around the world. There are 183 plots in all to carry out the ship recycling activities. Till today Alang has provided approximately 23 million tonnes of steel in the last 10 years. On 17-2-2006 when the above writ petition came up for hearing before this Court, we found the controversy concerning ship-breaking a recurring controversy. Therefore, this Court decided to lay down norms concerning infrastructure, capacity of Alang to handle large volume of ship-breaking activity, safeguards to be provided to the workers who were likely to face health-hazard on account of the incidence of ship-breaking activity, the environmental impact assessment, regulation of the said activity and strict regulation of the said activity. Accordingly, this Court constituted a Committee of Technical Experts to submit a report on the aforestated aspects. This case is related to dismantling of ships. Dispute relates to whether permission should be granted for dismantling of "Blue Lady" ship. Appellant alleged that ship contained toxic wastes like asbestos and also contained radioactive material on board. Appellant also raised objection with regard to non-quantification of two contaminants ACM and PCB. Atomic Energy Regulatory Board had certified that vessel contained no more any radio active material in the ship. A Technically Expert Committee was constituted in which this Court directed inclusion of Gujarat Maritime Board (GMB) and Gujarat Pollution Control Board (GPCB) to assist this Court. A Report was submitted by Technically Expert Committee (TEC) in which apart from examining the general compliance of the conditions, specific focus had to be drawn on three particular aspects, namely, (a) whether pre-conditions for dismantling have been complied with; (b) whether 80 per cent of the asbestos is reusable as is contended by the applicant; (c) what steps have been taken to control the environmental impact of asbestos dust generated in the process of dismantling. The report clarifies that "beaching" is an irreversible process. TEC has also examined the recycling plan and the dismantling plan submitted by the recycler. Apart from the GMB and GPCB, various other authorities like Gujarat Enviro Protection and Infrastructure Ltd. (GEPIL) have also contributed their knowledge and expertise in the preparation of the report dated 10-5-2007. There was also an apprehension rightly expressed by the petitioner regarding radioactive material on board the vessel "Blue Lady". Therefore, an immediate inspection of the said vessel beached at Alang since 16-8-2006 was undertaken by Atomic Energy Regulatory Board (AERB) and by GMB. The apprehension expressed by the petitioner was right. However, as the matter stands today, AERB and GMB have certified that the said vessel Blue Lady beached in Alang no more contains any radioactive material on board the ship. At this stage, it needs to be mentioned that breaking of the vessel Blue Lady will provide to this country 41000 MT of steel and it would give employment to 700 workmen. One of the main objections raised on behalf of the petitioner was regarding non-quantification of two contaminants, namely, ACM and PCB (rubber). ACM exists as material of construction in various vessel components like partitions, walls, and ceilings etc. which are an integral part of the vessel structure. In the present case, the vessel does not contain single kilogram of asbestos and/or ACM as cargo. However, there is presence of ACM as "material of construction" in various vessel components like partitions, walls, ceiling etc. Major quantity of ACM (85%) is in the partitions and ceilings of rooms and galleries. These ACMs are in the panels. TEC, in its report, stated that recyclable material alone can be sold; therefore, PCB cannot be sold. One of the main objections raised on behalf of the petitioner was regarding non-quantification of two contaminants, namely, ACM and PCB (rubber). At this stage, may be clarified that ACM exists as material of construction in various vessel components like partitions, walls, ceilings etc. which are an integral part of the vessel structure. In the present case, the vessel does not contain single kilogram of asbestos and/or ACM as cargo. However, there is presence of ACM as "material of construction" in various vessel components like partitions, walls, ceiling etc. Major quantity of ACM (85%) is in the partitions and ceilings of rooms and galleries. These ACMs are in the panels. TEC, in its report, stated that recyclable material alone can be sold, therefore, PCB cannot be sold. The report further indicates that the quantity of PCB in the present case has to be dumped in land-fills. It is important to note that there are only two alternatives, namely, incineration or dumping PCB in land-fills. In both the cases, there is likelihood of pollution. As regards asbestos is concerned, we find that 85 per cent is insulation and panels. This is where the dismantling plan has to be applied. It is this plan which takes care of the panels and insulation containing asbestos. Under Section 3.3.2, the recycler was required to submit a dismantling plan containing requirements to be complied with. Report of TEC showed that dismantling plan submitted by recycler complied with Section 3.3.2. Removal plan submitted by recycler for removal of ACM and PCB had been approved by TEC. Similarly, waste generation was also done as suggested by TEC storage of contaminated wastes in the land-fills has also been incorporated as per recommendations of TEC. 85 per cent of asbestos were reusable and dismantling plan complied with Section 3.3.2. Air monitoring has to be conducted for the air-borne ACM, if any. For that purpose the Committee has recommended appropriate respiratory protection to be provided to the workmen. For each category of work to be done in different areas of the vessel, gears have been provided to the workmen in the form of whole body coveralls, gloves, safety shoes, helmet, safety goggles etc. Similarly, as regards waste generation, the TEC report suggests by way of protection air monitoring respiratory protection to employees, leak tests, negative pressure checks etc. Similarly, storage of contaminated wastes in the land-fills has also been incorporated in the recommendations of the TEC. Therefore, in our view, in the light of the above conditions to be fulfilled by the recycler had satisfied all the conditions and principle of sustainable development based on the concept of balance stands satisfied. Thus permission was granted to the recycler to dismantle the said ship.

Another eye opening case related to the Environmental Law is **Parthiban Blue Metal** v. **The Member Secretary (Tamil Nadu Pollution Control Board) and Others.**[18] This was an environment case dealing with Crushing Units in the Residential Area of Trisoolam Village, Kanjipuram District, Tamil Nadu. Notice was issued to appellants requiring shifting of units. Appellant replied that units functioning in non-urban zone and according to the appellants they had obtained "no objection certificate" from the Tehsildar, Divisional Fire Officer and the Panchayat Union for the purpose of running the units. After Water (Prevention and Control of Pollution) Act, 1974 and the Air (Prevention and Control of Pollution) Act, 1981 came into force, appellants applied for consent from the concerned authorities under these Acts. But the Principal Secretary, Tamil Nadu Pollution Control Board, issued notice to each of the appellants requiring shifting to alternative sites since the units were located at about 325 metres from the residential colony. Due to filing of writ petition by several welfare associations, District Environmental Officer (DEO) issued another notice observing that units located within 500 metres from residential area requiring penal action. DEO's notice was challenged before High Court but High Court dismissed writ petitions. Hence, in present appeal, after examining evidences and circumstance Court observed that factual position was not examined by High Court and thus matter was remitted back to High Court for reconsideration—Parties permitted to place material in support and appeal disposed of.

Government of Karnataka and Others v. Smt. Gowramma and Others.[19] This was the case dealing with Karnataka Preservation of Trees Act, 1976; Karnataka Forest Act, 1963. In this case the plaintiffs and their predecessor had felled down silver wood, jungle wood and other varieties of trees in the scheduled land by spending lot of money and had cultivated the said land with coffee crop. In order to regulate the shade in the scheduled property and also for cutting and felling of silver wood, jungle wood and other trees, the plaintiffs had applied for permission for cutting and felling of the silver wood, jungle wood and other trees. Before granting the felling permission of the said trees, a joint survey was carried out by the forest authorities as well as the revenue surveyors. Thereafter the plaintiff was granted permission for cutting and felling of the silver wood, jungle wood and other trees. The plaintiff cut and felled down the trees. Now the plaintiff was entitled to the value of timber@ Rs. 125 per CFT and at Rs. 150 per CFT at prevailing rates. Defendants took the stand that the permission was conditional and was never challenged. The trial judge dismissed the suit, holding that there was no question of the plaintiff's making the claim for value of timber transported. Now an appeal was filed before High Court for granting relief to the plaintiffs, i.e. the present respondents, reliance was placed on certain judgements of the High Court where it was held that in respect of reserved trees, the ownership was not with the government but was with the owner of the land. Accordingly the appeal was allowed. In support of the appeal, learned counsel for the appellant-state submitted that the grant of permission was governed by the

Karnataka Preservation of Trees Act, 1976. Permission is required for the felling of trees irrespective of whether they are situated in private or in governmental land. Hence, there was no challenge to the conditions stipulated and it was accepted that the trees were reserved trees. What is the effect of this admission was not examined by the High Court. Therefore, looked from any angle, the judgment of the High Court is clearly unsustainable and is set aside. The appeal is allowed but without any order as to costs.

Bhag Singh and Others v. **State of Punjab and Others**[20]. A Town Planning case involving violation of Master Plan. The case involves violation of—Air (Prevention and Control of Pollution) Act, 1981—A writ was filed seeking closure of industries operated by non-official residents in residential area. Respondents resisted petition on ground that factories were operational as per the provision of Old Master Plan whereby it was declared industrial area. Board on being directed submitted report on action taken under Act—Petition disposed with direction to verify pollution norms and standards periodically. An appeal was made in which appellants contended that as per new Master Plan properties were earmarked as residential area; hence impugned judgment was not in order. Respondents resisted appeal on ground that new Master Plan was not applicable as it is yet to come into effect. They further held, since new Master Plan was yet to be notified its effect thereof could not be considered at this juncture. Further more they stated at the time of filing of writ old Master Plan was in effect and if in case of there being any change introduced by any subsequent Master Plan its effect thereof can be considered in terms of Notification of said Master Plan. Hence, the appeal was dismissed.

Sheikh Ikram, Sheikh Israil and Others v. **State of Maharashtra and Others**[21], Appellants were residents of Bhandara involved in the business of manufacturing brass utensils. The area being densely populated the noise was creating extra noise pollution in the vicinity and also there was a school nearby. Notice was issued by the superintendent of police whereby appellants directed to stop their business as activity causing noise pollution affecting neighbours, teachers and students at around nearby houses of the appellants. The defendants took stand that their business was pre-existing before the opening of the school. Report was called for and filed by State Pollution Board suggesting that high noise pollution amounting to nuisance. The concerned authority has been empowered by State Government to prohibit the level of pollution. Similarly, earlier a writ petition was filed in the Nagpur Bench of the High Court. In that case applications for interventions were filed on behalf of the school and some local residents. The writ petition was disposed of granting liberty to the parties to place the materials in support of their respective stands before the concerned authorities. The Noise Pollution (Regulation and Control) Rules, 2000 have been framed in exercise of powers conferred by Clause (a) of Section 25 of the Environment Protection Act, 1986.

Table 12.1: The Government ambient air quality standards in respect of Noise Pollution

Area Code	*Category of Area/Zone*	*Limit in db at Day Time*	*Limit in db at Night Time*
A	Industrial Area	75	70
B	Commercial Area	65	55
C	Residential Area	55	45
D	Silence Zone	50	40

They pleaded for an opportunity to reduce the noise level and remedial measures to be taken. The court declared, in the circumstances, we direct that the appellants are permitted to give a concrete proposal as to how they shall ensure sticking of the norms within two months. The proposal shall be dealt with a decision to be taken within three months. The appellants may, if so advised, and as contended move the authorities for making available alternative site. The feasibility by such a request shall be duly considered by the authorities.

T.N. Godavaraman Thirumulpad v. **Union of India (UOI) and Others**[22]. The case dealt with Orissa Forest Act, 1972. The case refers to attack on forest cover areas of Orissa where M/s Vedanta Aluminium Ltd. has filed an application seeking clearance of proposal of use of 723.343ha of land (including 58.943ha of reserve forest) in Langigarh Tahsil of Kalahandi District for setting of Alumina Refinery. The decision involves considering how much damage to the environment and ecology will the proposed project have. The project also includes mining of Bauxite deposits at top of Niyamgiri hills. The estimated Rs. 4000 crore project depends heavily on mining.

Whereas ecologists argue mining would deeply destroy flaura and fauna of the region. Niyamgiri hills are vital wildlife habitat, part of it constitutes, elephant corridor and hills are also a important water source of two rivers. According to CEC use of forest land in an ecologically sensitive area should not be permitted. But on the other side we have picture of abject poverty in which local of Langigarh Tahsil include tribal people. There is no proper housing, no hospitals, no schools and poor living conditions. So keeping the above aspects in mind and keeping in consideration of 8-9 per cent GDP growing rate the court though of balancing development *vis-à-vis* protection of environment and principle of sustainable development. The court decided that M/s. SIIL of which M/s. VAL is a subsidiary shall provide with permanent jobs to tribals and particularly land-losers. An affidavit was received regarding outsourcing of jobs to tribals by M/s VAL Ltd. The court held that if M/s. SIIL is agreeable to the aforestated Rehabilitation Package then they shall be at liberty to move this court by initiating a proper application. This court is not against the project in principle. It only seeks safeguards by which we are able to protect nature and sub-serve development. I.As, are disposed of accordingly. However, we once again reiterate that the applications filed by M/s. VAL stand dismissed.

Though objective acceptance of the decisions pronounced by judiciary cannot be claimed; but it is also tedious to deny that judiciary does adequate justice in case of environmental issues. It protects environment in many instances; no doubt it considers developmental concern also, wherever those are the need of the hour. But judiciary alone cannot handle the environmental issues, because it is a joint onus of other zones of administration also.

Administrative and Legislative Measures for Environmental Protection

Environmental degradation in form of deforestation, water depletion, drought, soil erosion, etc., has caught whole mankind into its prey. The problem has transcended local as well as national boundaries and reached an International pedestal now. Hence, it will be justified if 'epidemic' term is used to refer to environmental deterioration. It is affecting every individual, all species and the geology of the world, which infract reinforces intense scarcity of natural resources. Solution of this issue needs to involve all communities, governmental incumbents, institutions, non-state actors and common masses too. In the following part of the paper, we will discuss some possibly practical solutions to curb the problem of Environmental Degradation. Some of these steps have already been initiated, in context of which strict implementation needs to be made, so as to bring about the expected changes.

To check unregulated use of environmental resources, the application and implementation of Environmental Impact Assessment (EIA) programme is an important concern. Major feature of this programme is that it involves search for the methods to manage environment in compatibility with the complex world of today. Such an attempt will help save the nature from ever increasing materialistic and exploitative tendencies of men and defend the environment from destructive and inadvertent activities of human beings to develop the economy (Kumar, 2002).

Another helpful instrument which can pave way for immense environmental improvement and modification is 'Environment Cost Benefits Analysis', which is used when a purposive project is launched. Traditionally, this cost benefit assessment used to be expressed only in economic terms, such as in terms of labour interests, borrowed capital, etc. But all this has not produced those desired results, which the community actually expected. If sustainable development is to be brought, then the costs of benefits associated with the projects and environmental impacts must be taken into account. This can make it convenient to predict any environmental changes expected to be incurred as a consequence of operating a project, in lieu of which a pre-hand precaution for an efficient management can be taken. This approach will be helpful to take up qualitative ramifications (Lee, 1985).

Pollution control has assumed a significant role and it commensurates with industrial expansion. A strong regulatory mechanism is required to ensure minimum environmental standards to achieve this objective. A total number of 2,155 pre- and post-1991 units of 17 categories of highly polluting industries have been identified for this purpose by Central Pollution Control Board. The State-wise summary status of these 2,155 units is given in Table 12.2. Out of these units, 1,877 units have provided adequate facilities to comply with the standards, 225 units have closed down and the remaining 53 units do not have adequate facilities to comply with the standards. Legal actions have been initiated against all the defaulting units.

Table 12.2: Statewise Summary Status in 17 Categories of Highly Polluting Industries

	State/UTs	*Total No.*	*Status (No. of Units)*		
			Closed	*Compliant*	*Defaulters*
1.	Andhra Pradesh	269	29	240	0
2.	Arunachal Pradesh	0	0	0	0
3.	Assam	16	3	12	1
4.	Bihar	40	19	27	0
5.	Chhattisgarh	25	2	21	2
6.	Goa	8	0	8	0
7.	Gujarat	283	10	272	1
8.	Haryana	107	24	69	14
9.	Himachal Pradesh	11	0	11	0
10.	Jammu and Kashmir	10	3	7	0
11.	Jharkhand	21	3	16	2
12.	Karnataka	116	17	99	0
13.	Kerala	43	6	37	0
14.	Madhya Pradesh	78	15	61	2
15.	Maharashtra	392	26	356	10
16.	Manipur	0	0	0	0
17.	Meghalaya	1	0	1	0
18.	Mizoram	0	0	0	0
19.	Nagaland	0	0	0	0
20.	Orissa	29	3	21	5
21.	Punjab	72	9	60	3
22.	Rajasthan	108	8	96	4
23.	Sikkim	1	0	1	0
24.	Tamil Nadu	156	2	154	0
25.	Tripura	0	0	0	0
26.	Andaman and Nicobar	0	0	0	0
27.	Chandigarh	1	0	1	0
28.	Daman and Diu, Dadra and Nagar Haveli	0	0	0	0
29.	Delhi	5	1	4	0
30.	Lakshwadeep	0	0	0	0
31.	Pondicherry	8	1	7	0
32.	Uttarakhand	20	0	20	0
33.	Uttar Pradesh	263	27	232	4
34.	West Bengal	66	17	44	5
	Total	**2155**	**225**	**1877**	**53**

Source: Ministry of Environment and Forests, Economic Survey, 2005.

In order to strengthen the environmental management capacity at National, State and local level, the Ministry of Environment and Forest had launched an Environmental Management Capacity Building Technical Assistance Project in 1997 with assistance from the International Development Agency. This project has an outlay of Rs. 140.62 crores. The areas identified are capacity environmental administration, decentralization of environmental management, implementation of environmental laws, and monitoring compliance in specific priority areas (Economic Survey, 2004).

At the same time, industrial growth has been at below 10 per cent for several years now. One of the most important challenges in Indian economic policy consists of devising strategies for obtaining industrial growth in excess of 10 per cent, which infact bestows deterioration to the national environment.

It is well interpreted that industries pollute the environment. But still, industrialization cannot be barred because of being a germane core of economic development of the nation. A recommendation can still be enlisted; that the more polluting units more be substituted in their structure and functioning as well.

The Non-judicial Responsibility and Environmental Protection

The Non-Judicial responsibility refers to the 'out of court' shoulders. These shoulders are of the ordinary masses functioning in formal, semi-formal and even informal setups. Community participation is one of the efficacious factors having potential to function in a productive manner in environmental management. In the most contemporary interpretation, community participation is a tool for strengthening democracy. But this participation must go on systematically not randomly. It must bifurcate the contributions into two parts; mainly amenable to community management and to global management of environment (Aggarwal and Dubey, 2002). One of such successful step was United States National Environmental Policy Acts (NEPA), 1969. It aimed at involving the government, industry, community at large, but India has never used it as a milestone in the management paradigm. The results can be the most desirous and expected ones if the public which suffers, plays a large foot of role in addressal and redressal of its own grievances. This needs a moral boost up by the government.

In such a situation, the role of community in the redressal of its own suffrage takes immense importance. Legislature, executive and judiciary get encapsulated in the construct of the social order itself because the social order is excessively powerful. This manually constructed order, endowed with an intellectual and phenomenal ability is so strong that it turns the direction of the most formal units functioning in it vicinity. Civil society is also such an instrument, which abridges the gap between the environmental targets and the success desired thereon. The role of 'Civil Society' holds an ocular prominence in the contemporary era, where the quantitative increase in the functions of the governance is not being accomplished with a desirable qualitative performance of the same. Civil society pools efforts in improvement paradigm besides paving a way to extract the most practical solutions to an issue after observing the existing problematic scenario in a grounded way. As per the views given by Brabant (2006), the governance and the role of civil society in achieving good governance have received much attention in mainstream development thinking in the 1990s. In addition, he propounds that Governance and Civil society have also entered in the humanitarian sector. The post-conflict reconstruction agenda has expanded from the rehabilitation of physical and social infrastructure to, ambitiously, the reconstruction of viable societies with power-sharing politics and an assertive civil society. This power of assertion of the civil society attains a dominantly emphatic form in the course of improvement of the undertaken structure. Hence, an active civil society is seen as an important basic framework and precondition to reduce the intensity of the social and other issues, which comprises environmental degradation.

Moreover, the fast developing nations like India still stand confused in a simple question, "which out of the two, Judicial Activism and Social Activism, is paramount"? The reason behind this perplexity is not the weakness of judiciary rather it is the power of the emphatic 'push' composed by the most ordinary masses; which further influences the structure and the functions of judiciary and the other units of the governance. So, the cases discussed in the paper; somewhere or the other reflects a strong component of community as well as individual consciousness about the environment. This evidently establishes a notion that Environmental Law and the related legislations is not an end in them, rather they are the means to achieve the ends as desired by the communities. Environmental Law has to be interpreted, applied and amended in terms of 'community needs'; which is further an entangled concept of socio-cultural and economic domain of the variable human societies.

The past-experiences have proved it credulously that the volunteering groups have overpowered the formal structures and have even performed better than those structures tied in formalism. The decorous structures like Judiciary, the burnishing performance of the Legislature and the blatant Executive, many a times corroborate with the instruments thrown forth by the ordinary masses; as these bodies do not deny that the victims of an issue, are the most efficacious source of extracting the most practical solutions to curb it. Hence, the non-judicial responsibility can perform in a way class apart; the domain of work may be any! Expecially, in case of environmental issues, the community participation is a boon. The communities dependent upon environment are emotionally attached with ecological units, which makes them respond more sincerely to the improvement efforts, in case if environment moves towards degeneration.

Conclusion

Environment is such an aspect of human life, which influences various choices of social survival. Even economy and environment have a mutual dependability. The environmental frontier determines the growth of the economy and the *vice versa* is also applicable. Alongside, legislative and judicial attempts made by Indian Government incessantly, have been altering the existing paradigm to bring about desirable changes. The cases surveyed in this Annual Survey show that the courts have an absolute considerations about Sustainable Development and aim at an absolutely functional equity between environmental and economical concerns. If 'Sustainable Development' is to be achieved, it is a prerequisite to create a balance between procedures adopted for economic development and environmental maintenance. Preservation and protection of environment and the environmental units is not a sole responsibility of the Government or administration, rather it is a joint liability of civil society and the base of civil society is common man's participation. If this participation is assisted and encouraged by judiciary, it inspires such protective action in future, which is a dire need of the modern world featured by developmental complex matrix. Hence, to strengthen judiciary, by way of providing it subject-matter to function in environmental context, it is important that the common masses contribute in terms of 'human and ideological capital'.

NOTES

1. (1996) 3 SCC 212
2. (1996) 5 SCC 647 at 659-660
3. (1997) 5 SCC 10
4. AIR 2006 SC 2893.
5. AIR 2006 SCW 2121.
6. AIR 2006. SCW 1309.
7. AIR 2006 SCW 1392.
8. AIR 2006 SC 2038.
9. AIR 2006 SC 1350.
10. AIR 2006 SC 1428.
11. AIR 2006 GUJ 97.
12. AIR 2006 Calcutta 75.
13. AIR 2006 All 246.
14. AIR 2006 Jhr 96.
15. AIR 2006 AP 386.
16. AIR 2006 SC 1774.
17. MANU/SC/7894/2007.
18. MANU/SC/7060/2007.
19. MANU/SC/2874/2007.
20. MANU/SC/7003/2007.
21. MANU/SC/1839/2007.
22. MANU/SC/1279/2007.

REFERENCES

Aggarwal, Manmohan, *Globalization and Millennium Development*, Social Science Publication, New Delhi, 2007.

Aggarwal, and Dubey, *Environmental Controversies*, APH Publishing Company, New Delhi, 2002.

Brabant, Van K. "Civil Society and Substantive Democracy: Governance and the State of Law in Belgium", in *Development and Social Action* by M. Kothari, Rawat Publishers, New Delhi, 2006.

Chahal, S.K., *Environment and World Life: Towards a New Paradigm,* Ashish Publishing House, New Delhi, 1994.

Economic Survey, Ministry of Finance Publication, Government of India, 2004.

Ghosh, A. *Urban Environment Management: Local Government and Community Action,* Concept Publishing Company, New Delhi, 2003.

Human Development Report 2006, United Nations Development Program Publication, UN Plaza, New York. 2006.

Jaswal, P.S., *Environmental Laws,* Pioneer Publications, New Delhi, 2006.

Kumar, H.D., *Sustainable Human Ecology,* Atlantic Publishers, New Delhi, 2002.

Lee, A. James, *Environment, Public Health and Human Ecology,* John Hopkins University Press, London, 1985.

Robertson, R., *Globalization: Social Theory and Global Culture,* Sage Publishers, London, 1992.

Speth, L.J. *Global Environmental Challenges: Frontiers to a Sustainable World,* Orient Longman, New Delhi, 2006.

13

Environmental Protection Vis-a-Vis Judicial Activism

PRASHISH KANWAR and CHANDAN SINGLA

The black ebony staves of judiciary which has thumped time and again for protection of man miniature against excruciating blows of evil is known on the aspiration for protecting environment. Judiciary is having a coherent vision on the environmental protection. However, the problem of law making and amending is really onerous in this area. As there are certain things like industrialization, urbanization, sundry cultural and moral values of humanity those hamper or create a razzmatazz of legal norms which are really hard to be deciphered out. In today's emerging jurisprudence, environmental rights incorporates of collective rights are described as 'third generation' rights. The "first generation rights" are generally political rights while "second generation" rights are called socio-economic rights as found in the international convert on economic, social and cultural rights[1]. There is a prominent saying "The times have changed and you must too unless the times wont forgive you" so according to the changing trends of the society from time to time, law also has to evolve accordingly. Earlier, there were many human activists who worked for freeing the society from the existing problems. But in the fast moving world of today, where a person hardly finds time off for his family and close ones, it is a disappointing fact that there is no one to look after the matters of public importance. Hence, fields like environment protection go unnoticed. For this PIL has emerged as a Midas touch and is proving to be very effective. However the role of judiciary is really important as the role of mitochondria of living human sell. Had judiciary turned the deaf ear towards environmental problems it could not be in anyway came to celluloid. One significant fact to support the sensibility of judiciary is the case of Subhash Kumar *v.* State of Bihar[2] wherein personnel grudges of two parties the judiciary put life in the cold letters of Constitution, i.e. the environmental protection which previously was a fundamental duty under Article 51(A) also came as fundamental right under Article 21 of the Constitution of India. No matter how much criticized it is, nc matter how much unidentified it is but one thing to which everyone take leave to doubt is the massive contribution to the welfare of the environment.

Development versus **Pollution**

It is the conscience of the judges which plays the paramount role i the decisions regarding the environmental decisions. Moreover it i not a hard and fast rule that in environmental litigation the momentun will be on the side of the party who is opposing the spread o industrialization or urbanization but the endeavour of the judicia conscience will remain on the hauling out welfare of the public—A the soaring leaps of industrialization cannot be put at a halt blindl as not only it amounts to the stagnancy in the progress of the countr however it also inculcate the new versions of litigation through th shoulder of "environment fortification justice". Elaborating it furthe the conscience of the judges has to be pass through sharp edge weapo of iconoclasts so that their decision does not in any way defeat th provisions that are meant as the edifice of all the other legal norm Where prayers were recited in church by using loudspeaker drun etc. there was a complaint of noise pollution. Directions was give by High Court to authorities to follow guidelines for control of nois pollution as issued in 1995 AIHC 4168 and to make church to kee speakers at lower level. Held that such directions by High Court we not illegal in view of clear pollution control provisions. Appreciab our judiciary is imbibing rather welcoming the international pollutic control principles. Compensation to be awarded must have some bro correlation not only with the magnitude and the capacity of t enterprise but also with the harm caused by it. This *pro bono publi* judicial process has three main features. Firstly, the judges a insulated towards political pulls and pressures so they give decisio according to best of their knowledge and checking everything on anvil of public welfare. Second feature of the judicial process is that it frequently compels the defendant project is challenged in court by an environmental group, the promoters frequently spend more funds on environmental impact appraisal and pollution abatement, than they otherwise might. Third asserts that it should be remembered that the courts can only respond to the cases that come before them[3]. Courts of law in India are taking really active part in conservation of environment as there is a famous case *Ganga Pollution Case*[4] where massive judicial effort was taken to curb environmental problem by cleaning Ganga river. So there is every reason to believe that judiciary has played a role of conscience keeper.

Messenger to 'providers of the justice'

The PIL is becoming the salvage to the poor and suppressed people and an instrument by which the ears of judiciary could be lend quite easily. "The role of the telegram to the door of justice is very well played by one post card which was never before played that expediently by more than one smidgens of paper". Rural litigation case is said to be the harbinger of the revolution which has opened the floodgates of the PIL's for curbing environmental problems. The filing of petition is not the only thing rather after filing the petition these are the judges who are to entertain petition and that typically depends on the bent of mind of the judges. The cases raising questions of the environmental degradation were really speaking cases against inaction of the State or wrong action of the State. The Court made it clear that petition alleging environmental pollution caused by private industrial units as against the Union of India, the State Government, and the Pollution Control Boards established under Environmental Protection Act, 1986, which were supposed to prevent environmental hazards. Their failure to perform their statuary duties resulted in violation of the right of the residents to life and liberty guaranteed by Article 21 of the Constitution. The Court entertained a petition by the residents of Banglore objecting to the approval of development schemes that was likely to adversely affect the quantity and quality of water of a river. The Court had to actually monitor restrictions on mining operations that were hazardous to the health of the people living in surrounding areas by appointing a committee to oversee the implementation of the courts directions. The Court has dealt with environmental issues such as pollution by tannery industries protection and conservation of forests urban and solid waste management vehicular pollution in Delhi due to location of mechanized slaughterhouses and protection and conservation of wildlife. The Supreme Court was also approach against degradation of the Taj Mahal, pollution of the river Ganges by Calcutta tanneries that discharged untreated noxious and poisonous effluents into it, and protection of the people from stone quarrying in the Dehradun region. The Court also laid down the principle that the polluter pays for the pollution. In another case, the Court held that in matters of environment, the burden of proof will lie on the party that wants to change the *status quo*[5]. As the *lassiez faire* is turning to the welfare state so there is high time when somebody has to be the crusader or be the defender against the challenges in the society. Judiciary through PIL's is very well performing the duty. There are oodles of problems those are accumulating in front of us as the concept of environmental protection is a recent born practice in judicial parlance. Moreover the action of the administrative agencies cannot be taken in a lighter way. Judiciary has to concentrate on each and every action of administration in protection of environment which no denying the fact is considered as judicial activism but there is no harm in taking active cognizance of environment protection cases.

Articles 226 and 32 of the Constitution of India are the two eyes of judiciary through which whole Indian territory could be put under vision by which our judicial system becomes a proactive and Janus-faced. The invocation of Articles 226 and 32 is becoming so prominent because of its expedient nature instead of conventional suit which time and again labelled as "justice delayed is justice denied" which can be replaced as saying "decision delayed is decision denied" as there is no surety of justice thereafter. Furthermore the principles laid down in Bhopal gas case, the Supreme Court formulated the doctrine of absolute liability for harm caused by hazardous and inherently dangerous industries by interpreting the scope of the power under Article 32 to issue directions or orders which ever may be appropriate in appropriate proceedings. According to court of law the power could be utilized for going new remedies and fashioning new strategies[6]. The Bhopal gas tragedy is not considered as one case which was more than a apocalypse at catastrophic level rather it was that big disaster due to repercussions of which the souls of the died persons are still hanging in the air for the justice and the justice is not confined to deciding that very case rather for making for framing such a laws which are not only for the decisional benefit of the case, but for avoiding the upcoming Bhopals. The directions were given by Court for disciplining the development processes, keeping in view the demands of ecological security and integrity[7]. In one of the earlier cases rural litigation case that posed an environment development dilemma Supreme Court gave directions that were necessary to avert an ecological imbalance such as constitution of expert committees to study and suggest solutions establishments of a monitoring committee to oversee a forestation programmes and stoppage of mining operations that had an adverse impact on ecology. Writs as Remedies in Environmental cases—

- The aggrieved individual or
- By way of '*Public Interest Litigation*'.

The principle public law remedy is that of writs. But it should not be overlooked that our law of civil procedure contains specific to enable two or more persons having a legitimate interest in the subject matter to seek remedy through court for

- Remedying public wrong (Section 91, CPC)
- Remedying breaches of public trusts (Section 92, CPC)

It is indubitable that the courts of law are having very generous demeanour regarding the acceptance of PIL whenever there is question of degradation of environment. However, the judicial conscience in scrutinizing the PILs is becoming investigating and operational. Furthermore judiciary is becoming much attentive than ever before in eying the PIL's coherently so that some amount of obstruction could be put to the floodgates for the PILs as in famous case of *Subash Kumar* vs. *State of Bihar*[8] the Supreme Court upheld that affected persons or even a group of social persons or generalist but not at the instance of persons who had a bias or personnel grudge or enmity could initiate PIL for environmental rights. So, there is every reason to believe that the relieving the concept of *locus standi* and emerging PIL has revolutionized the process of environmental protection.

Bench and Bar in Environmental Protection

One thing which is uncurtain is the role of the Bar in the environmental protection. There is no denying the fact that it is the words of the judges which holds the paramount importance however it is an axiomatic truth that rationality and they clear the just over the facts for proper litigation. They are colossus in accentuating the pronouncements by judges and removing the dust from the anticipatory visions. M.C. Mehta one of the world's most renowned and successful activists. A lawyer by profession and a committed environmentalist by choice, he has made the fight to protect India's environment his unending mission. He has pioneered legal activism for environmental protection and is proof that one man can make a difference[9].The lawyers are salvage to the degrading environment but also nightmare to those who are encroaching the environment, i.e. leaping towards industrialization and urbanization. Mr. Mehta is indubitably a public spirited lawyer time and again has set glorious precedents, which actually has opened a way for the green coloured pieces of paper i.e. PIL related to environmental problem to enter the dwellings of law. Furthermore, Mr. Mehta who should be called as Mr. Green, in a case *M.C. Mehta* vs. *Union of India*[10] concerned closing down a chlorine plant of Shriram industries. The one known as Oleum gas leak case was decided by five judge's constitutional bench of Supreme Court, in view of gravity of the legal questions under considerations. The court of law has welcomed this green letter of M.C. Mehta with open arms and imposed absolute liability on the hazardous industry. In the next case there is case again *M.C. Mehta* vs. *Union of India*[11] in which the contemplation was done regarding discharge of effluents by tanneries and chemical industries into river Ganga. The Supreme Court ordered its office to serve notice about institution of that case to all such industries and after hearing both sides ordered those tanneries not having pretreatment plants approved by Pollution Control Board to stop their discharge of trade effluents. In yet another of Mehta series, the issue was whether the petitioner who was not a riparian owner could be granted standing to move for prevention of nuisance for pollution in river Ganga. The Court held that he was a person interested in protecting the lives of the people who make use of the waters of the river and so could move. In the Courts view, in public interest litigation, it is reasonable to allow any person to take proceedings on behalf of the community at large. So the attempts of M.C. Mehta has turned into anxiety which is really appreciable and the works of justice Kuldip Singh cannot be ignored but the point of worry or the question to be answered is who is another M.C. Mehta and Kuldip Singh. The activist working for environment are not always genuine that is why the PIL is criticized as cosmetic activism.

National as well as International Endeavours

International treaties and conventions have laid down many norms. The first was the United Nations in its International Conference[12] in the year 1972, laid down its agenda as "*to defend and improve the human environment for a present and future generation has become an imperative goal for mankind*". It called upon all the Governments and people of the world to exert common efforts for the preservation and improvement of the Human Environment. This Conference at Stockholm became the turning point for the Environmental Jurisprudence. It laid down twenty-six principles, which are known as *Magna Carta* on Human Environment. The principles are categorized under the various heads:

(a) two proclamation rights[13];
(b) four concerning the conservation of resources[14];
(c) two on pollution[15];
(d) eight on developmental issues[16];
(e) nine on specific non-legal topics[17]; and
(f) one on state responsibility[18].

Then there is the General Assembly Resolution in 1972[19], emphasizing the need for the active cooperation among the States in the field of human environment. Then the Charter of Economic Rights and Duties of States, in 1974, which put its emphasis on the protection, preservation and the enhancement of the environment for the present and the future generation. It also takes in account that the States should ensure that the activities within their jurisdiction or control do not cause damage to the environment of other States. Thus the awareness started to build up and many bilateral and multilateral treaties were signed between States for the cause of environment protection.

The paramount international effort at the preservation of the environment during the century after the Stockholm Conference is the Earth Summit[20]. The document produced at the Earth Summit has 40 chapters having 800 pages. In this Conference, 108 States participated. They together adopted three major agreements aimed at changing the traditional approach to development:

- Agenda 21—a comprehensive programme of action for global action in all areas of sustainable development;
- The Rio Declaration on Environment and Development—a series of principles defining the rights and responsibilities of States;
- The Statement of Forest Principles—a set of principles to underlie the sustainable management of forests worldwide.

The Earth Summit succeeded in presenting new perspectives on economic progress. It was lauded as the beginning of a new era and its success would be measured by the implementation of its agreements locally, nationally and internationally. The Earth Summit proved colossus amongst all subsequent UN conferences, which have examined the relationship between human rights, population, social development, women and human settlements and the need for environmentally sustainable development. Thereafter in the World Conference on Human Rights, held in Vienna in 1993, emphasized the right of people to a healthy environment and the right to development, controversial demands that had met with resistance from some Member States until the Earth Summit.

India was also one of the signatories of the Stockholm Conference. Therefore, in light of the promise made at the Conference, the Indian Parliament passed the 42nd Amendment to the Constitution in 1976[21] and incorporated specially two Articles relating to protection and improvement of the environment. Thus India became the first country in the world to leap forward for preventing this precious aura and have provisions in the Constitution.

Despite the provisions in Constitution of India, the Parliament has enacted various laws governing the protection and preservation of the environment. For example:

- Water (Prevention and Control of Pollution) Act, 1974.
- Air (Prevention and Control of Pollution) Act, 1981.
- Environment (Protection) Act, 1986.
- Forest Conservation Act, 1980.
- National Highways Act, 1983.
- The Drugs and Cosmetics Act, 1940.
- Motor Vehicle Act, 1988.
- Industries (Development and Regulation) Act, 1951.
- Wildlife Protection Act, 1972.
- Atomic Energy Act, 1962 etc.

In series of the endeavour there are more than 100 Centre and State Legislation providing different measures to control and maintain the safe environment in one or the other way. Even I.P.C.[22] and Cr.P.C.[23] contain various provisions for punishing those nefarious persons who causes damage to the environment threatening the existence of the mankind.

Excruciating Reality Through Legal Lens

All the enactments and norms established contain provisions for the protection of environment by imposing duty on every citizen as well on the State, but when these provisions are implemented human rights of individuals are violated in the name of environment protection. It is not always true to say that, the laws pertaining to protection of environment violates human rights but there are provisions under certain laws which when exercised abridges the rights of persons in exercising their free will. For instance, in the Forest Conservation Act, 1980, in which the Central Government the sole authority to give directions regarding the reserved forests and on such recommendations only the State Government can further make rules. By utilizing this power the State Government has made rules, which says that no person shall be allowed to enter into the forests or go out without the permission of the forest authorities. Similarly the rules again says that any property lying in the reserved area belongs to the Government and no other use of forest land or any portion of it can be made for any non-forest purpose[24]. Here the rule is enacted to protect the forest, but the question arises about the rights of persons who were residing in these forests. These tribal people were the protectors of these forests, the rights of the sons of the soil. By implementing these provisions the rights of these tribal people are infringed who were dependent on these forest for their livelihood.

In one pretext or the other most of these environment laws violate the rights of the people, the reason is that the Government in creating awareness or protecting the environment forgets about the human rights. These Acts hardly give any consideration to the living beings called human because their first concern was environment protection. For protecting and preserving the Human Rights the Courts have to interfere to provide appropriate remedy to the aggrieved, excavating their rights from Article 21[25].

The Supreme Court through a catena of decisions has opened many facets of Article 21 to protect the human rights with environment. In *Rural Litigation and Entitlement Kendra* vs. *State of Uttar Pradesh*[26] the right to clean environment as an integral part of right to life.

Other conflicting area where environmental lacks human face are:

(a) Under the right to life drinking water is one of the fundamental requirements but simultaneously State is obliged to preserve the groundwater or to restrict the use so that it can be prevented from being wasted.[27]

(b) State has an obligation to protect the forest but at the same time has to protect the tribal people whose livelihood is based on forest.

State is bound to close down those industries which are hazardous for environment, but at the same time has to keep in mind the livelihood of the people who work there. The Supreme Court from time to time has being issuing guidelines for the protection of Human Rights and Environment through various State laws. These series of repair work are coeval with violations as Supreme Court is actively taking cognizance of all such happenings. The norms[28] are:

- Every person enjoys the right to a wholesome environment, which is a facet of the right to life guaranteed under Article 21 of the Constitution.[29]
- Enforcement agencies are under an obligation to strictly enforce environmental laws.[30]
- Government agencies may not plead non-availability of funds, inadequacy of staff or other insufficiencies to justify the non-performance of their obligations under environmental laws.[31]
- The polluter pays principle which is a part of the basic environmental law of the land requires that a polluter bear the remedial or clean up costs as well as the amounts payable to compensate the victims of pollution.[32]
- The precautionary principle requires government authorities to anticipate, prevent and attack the cause of environmental pollution. This principle also imposes the onus of proof on the developer or industrialist to show that his or her action is environmentally benign.[33]
- Government development agencies charged with decision making ought to give due regard to ecological factors including:
  - the environmental policy of the Central and State Government;
  - the sustainable development and utilization of natural resources; and
  - the obligation of the present generation to preserve natural resources and pass on to future generations an environment as intact as the one we inherited from the previous generation.[34]
- Stringent action ought to be taken against contumacious defaulters and persons who carry on industrial or developmental activity for profit without regard to environmental laws.[35]
- The power conferred under an environmental statue may be exercised only to advance environmental protection and not for a purpose that would defeat the object of the law.[36]

- The State is the trustee of all natural resources, which are by nature meant for public use and enjoyment. The public at large is the beneficiary of the seashore, running waters, air, forests and ecologically fragile lands. These resources cannot be converted into private ownership.[37]

Thus the Supreme Court had taken into account the right to healthy environment along with the right to sustainable development and balanced them. This concept of right to healthy environment and sustainable development are the fundamental human rights implicit in the right to life, which has been constructed as such in many countries. The entire judicial construction by the Supreme Court and the High Courts also reveal the humanitarian approach to these environmental laws with the help of the public interests litigations. The Indian Supreme Court was the first to develop the concept of right to healthy environment as a part of life under Article 21 of our Constitution.[38] This principle is been now adopted and followed in various other countries now.

It is really flabbergasting that the law in its very way has encountered time and again the ditch which is itself created by law. elaborating it further the laws which are made to serve for the right of people are intersecting with the laws which are already existing.

NOTES

1. (See Vol. 25) 2000 *Columbia Journal of Environmental Law.*
2. AIR 1991 SC 420.
3. Thakur, K., *Environmental Law and Policy in India,* Deep and Deep Publication, New Delhi, 2005.
4. M.C. Mehta *vs.* Union of India AIR 1988 SC 1037 and 1115.
5. Speth, Gustave, James, *Global Environmental Challenge: Transitions to Sustainable World,* Orient Longman, 2005.
6. Tyagi, B.S., *Judicial Activism in India,* Srishthi Publication, New Delhi, 2000.
7. *Ibid.*
8. AIR 1991 SC 420.
9. M.C Mehta Environmental Foundation (MCMEF).
10. AIR 1987 SC 965.
11. 1992 SUPP (2) SCC 633.

12. UN International Conference on Human Environment at Stockholm, from 5th to 16th June, 1972.
13. Principles 1 and 21of *ibid.*
14. Principles 2 to 5 of *ibid.*
15. Principles 6 and 7 of *ibid.*
16. Principles 8 to 15 of *ibid.*
17. Principles 16, 20, 23 and 26 of *ibid.*
18. Principle 22 of *ibid.*
19. General Assembly's Resolution of 5th December 1972, designating June 5th as World Environmental Day.
20. United Nations Conference on Environment and Sustainable Development 1992 at Rio-de-Jeneiro.
21. The Constitution (42nd Amendment) Act, 1976 incorporated Articles 48-A and 51-A to the Indian Constitution.
22. Chapter XIV Sections 268 to 294A providing penal provisions to punish the person who pollutes the environment.
23. Chapter X Sections 133 to 146 and Chapter XI Section 144.
24. Section 2 of the Forest (Conservation) Act, 1980.
25. In the year 1984, the Supreme Court gave the judgment that Right to Life also includes right to pollution free environment in the case Bhandu Mukti Morcha *v.* Union of India 1984 (3) SCC 161.
26. AIR 1985 SC 652; See also, Subash Kumar v. State of Bihar, AIR 1991 SC 420.
27. The need for management of groundwater resources was recognized by the High Court of Kerala in a public interest litigation filed by the local islanders. Attakoya Thangal *v.* Union of India, 1990 (1) KLT 580.
28. Divan, Shyam, *Environmental Law and Policy in India,* Oxford University Press, New Delhi, 2007.
29. Subash Kumar *v.* State of Bihar, AIR 1991 SC 420. M.C. Mehta *v.* Union of India (Delhi Crushing Case), 1992 (3) SCC 256, and Virender Gaur *v.* State of Haryana, 1995 (2) SCC 577.
30. Indian Council for Enviro-Legal Action *v.* Union of India (CRZ Notification Case), 1996 (5) SCC 281, 294, 30131 Dr. B.L. Wadehra *v.* Union of India (Delhi Garbage Case), AIR 1996 SC 2969.
32. Indian Council for Enviro-Legal Action *v.* Union of India (Bichri case), IR 1996 SC1446, See also S.Jaganath *v.* Union of India (Shrimp Culture case), AIR 1997 SC 811.
33. Vellore Citizen' Welfare Forum *v.* Union of India, AIR1996 SC 2715.
34. State of Himachal Pradesh *v.* Ganesh Wood Products, AIR 1996 SC 149.
35. Indian Council for Enviro-Legal Action *v.* Union of India (Bichri case), AIR 1996 SC1446.
36. Bangalore Medical Trust *v.* B.S. Muddappa, AIR 1991 SC 1902.
37. M.C. Mehta *v.* Kamal Nath, 1997(1) SCC 388.
38. Bhandu Mukti Morcha *v.* Union of India, 1984 (3) SCC 161.

14

Role of Judiciary in Environmental Activism

RAJAN BANSAL and KARAN KALRA

Indian Environmental Law and Role of Judiciary

India has inherited a culture of tolerance, non-violence, equity and compassion for animate objects. In the olden times they were a part of daily life and synthesized with the religion. Religious teaching, social and political norms, economic policies treated man as a part of nature not as a moulder or superior to it. Air, water, land, animals and human beings are the creation of one superior power-God. Therefore, the fundamental ethic of behaviour was to live in harmony with each other, because it was well realized that each one of them is dependent on one another and destruction or damage to the other is the destruction of self and is complementary to each other. Without a natural and congenial environment human existence is not possible on earth. Since time immemorial the man had made conscious and determined efforts to make use of the natural resources and to modify his surroundings so that the adverse impact caused by extreme of temperature rainfall and predators may be reduced. Man always exploited the nature in greed of making his life more comfortable. Human activities create a large variety of wastes and by-products which accumulate over a period of time and may become toxic to the naturally growing plants, animals and the man kind. Like most of the developing countries, India too is facing environmental problems owing to the massive expansion of energy and resource-intensive industrial activities, urbanization, exploitation of forests, mining, energy-intensive agriculture and mega developmental projects. India being a country having large population is facing some typical environmental problems because of population load and requires special attention on issues like town planning and vehicular pollution. Indiscriminate use of chemical fertilizers and pesticides has added to the problem of environment. Many of the chemicals use for the proper growth of plants and animals have turned out to be harmful as their concentration rises; they contaminate soil, water and air resulting into the illumination of many species. Modern lifestyle has also given rise to the problem of greenhouse gases which are responsible for global warming, causing melting of snow at poles and thereby raising the level of sea endangering the terrestrial life. Man is really left with no choice than to fight seriously for the cause of betterment of existing environment and natural resources. It is high time to look into and examine the plans, schemes and projects having adverse effects on the environment so that a better, balanced and eco-friendly development ensuring pollution free air, potable water, and other elements of environment (Tiwari, 2006).

Conceptualization of the Term 'Environment'

The concept of environment is as old as the concept of the nature itself. It is a composite term referring to conditions in which organisms consisting of air, water, food, sunlight etc., thrive and become living sources of life for all the living and non-living beings including plant life. It can be understood in two dimensions, namely, in a narrow and broader sense. In its narrow sense environment denotes surroundings whereas in its broader sense it includes everything visible and invisible, living and non-living and their interaction in, on and above the surface of earth. According to K.R. Dikshit our minimum concern is the quality of space we live in, the purity of air we breathe, the food we eat, the water we drink and the resources we draw from our environment to support our economy. According to the United States Council on Environmental Quality, "environment" means man's total environmental system including not only the biosphere, but also his interactions with his natural and man-made surroundings (Padme, 2006).

Four decades ago, Environmental Defence helped launch the modern environmental era by winning a ban on DDT, the pesticide Rachel Carson warns about in *Silent Spring*. DDT causes eggshells to thin and break, threatening the survival of magnificent birds like the osprey, bald eagle and peregrine falcon. It is also a persistent poison that works its way up the food chain, thus endangering humans as well. The fledgling effort by a handful of scientists on Long Island to halt the use of DDT was a remarkable demonstration of how individuals can bring about lasting change. The group incorporated as the Environmental Defence Fund in 1967.

Our founders tried a novel approach, common today but unheard of in the late 1960s: the scientists teamed up with a lawyer and went to court on behalf of the environment. Their efforts led to a nation-wide ban on DDT and the birth of modern environmental law. The osprey has since made a dramatic recovery, and the bald eagle and peregrine falcon have been removed from the endangered species list (Tripathi, 2005). Soon we were hiring economists, engineers and eventually computer analysts to find ways to help the environment without harming the economy. In the process, Environmental Defence became one of America's most influential environmental advocacy groups, now with over 500,000 members and more Ph.D. scientists and economists on staff than any similar organization (Verma, 2003). From the beginning, we made a commitment not only to oppose ill-conceived policies, but also to propose alternatives. We still go to court when necessary, but increasingly we work directly with business, government and community groups, forging solutions that make sense for all.

The Present Indian Scenario

Environmental destruction and pollution has seriously threatened the human life, health and livelihood. The problem of environmental pollution is increasingly an world concerns; as it catches all countries irrespective of their size, level of development or ideology, as its prey. In India, as elsewhere in the world, uncontrolled growth and consequent environmental deterioration are fast assuming menacing proportions and all Indian cities and majority of the population are afflicted with the problem. Global warming, ozone depletion and toxic pollution are some of the negative effects of the existing development strategy.

Today most of our rivers are polluted. Deforestation in most of the regions is increasing day by day. Emission of poisonous gases and other harmful gases, liquids and disposal of solid wastes from the industries have almost become a regular phenomena of the present day. The problem of noise pollution, particularly in big cities is at alarming stage. Land erosion through winds and water has become a common feature. The growth of population in India has created several environmental problems such as overall reduction in agricultural land, lower productivity, deforestation, soil erosion, landslides, shortage of drinking water, increasing level of pollution, etc. Due to uncontrolled urbanization, environmental degradation has been occurring very rapidly and causing shortages of housing, worsening water quality, excessive air and noise pollution, dust and heat, and the problems of disposal of solid wastes and hazardous wastes. Urbanization, associated with industrialization has also created several environmental problems such as growth of slums, air and noise pollution by vehicular traffic, destruction of natural sites, vegetation, problem of waste water etc.

According to a World Bank estimate, India loses a whooping $80 billion annually on account of sickness and death from pollution and economic costs attributable to resource degradation. The loss due to poor environment is pegged in excess of $20 billion a year by conventional calculation and $80 billion in purchasing power parity terms, according to the World Bank annual environment review released in October 1998. According to the World Bank, deteriorating water quality due to poor sanitation, industrial effluents and pesticide run-off, lack of clean water, poor solid waste management and air pollution are the key to environmental problem in south Asia. However, the fundamental question before us today is whether we can allow the destruction of the environment leading to the destruction of all living creature including human beings on this planet! The answer is obviously-'No'. In the course of brutal exploitation of forests, indiscriminate quarrying, pollution of rivers and other water resources and a rapid increase of air and noise pollution; there is still a hope to protect the deteriorating environment through proper policies and management mechanisms. And the recent developments in the area of environment protection prove this point (Jaswal, 2006).

In order to achieve sustainable development 'environment protection' constitutes an integral part of developmental process; and it cannot be considered in isolation. Peace, development and environment are interdependent and indivisible when mechanisms are being designed for sustainability. Today, we are confronted with a perpetuation of disparities between and within nations; due to worsening of poverty, hunger, ill-health and illiteracy, and the continuing deterioration of the ecosystem on which we depend for our well-being. However, integration of environment and development concerns and greater attention to them will lead to fulfilment of basic needs, improved living standards for all; better protected and managed ecosystem and a safer and a longer future.

The problem of environmental pollution is the problem of developed, developing and underdeveloped nations. As far as the developing countries are concerned, "poverty and lack of development" constitute an essential element of the problem of environmental pollution. In fact, poverty is a potential cause of environmental pollution. The poor people who do not have means to get one meal a day, clothes to cover themselves and shelter to live in, cannot possibly think about the environmental protection. For them any method by which they can survive is the best, least caring about its effect on the environment. Other than pennry, over-production, nuclear radiations, over-exploitation of resources, industrial wastes in different forms, industrial accidents and the living style of the people are some of the contributing factors for environmental problem in the developed countries.

The present century, particularly in the later half has seen a lot of growth and economic development in almost all the countries. The methods of economic development which mankind has followed, are also creating environmental problem. With the industrial and technological development, mankind has certainly improved the economic conditions but has devalued the natural ecological in balance. Industrialization, urbanization and erosion of biodiversity have affected the natural environment adversely (Jaiswal, 2006).

Environmental Laws

In the Constitution of India it is clearly stated that it is the duty of the State to 'protect and improve the environment and to safeguard the forests and wildlife of the country'. It imposes a duty on every citizen 'to protect and improve the natural environment including forests, lakes, rivers, and wildlife'. Reference to the environment has also been made in the Directive Principles of State Policy as well as the Fundamental Rights. The Department of Environment was established in India in 1980 to ensure a healthy environment for the country. This later became the Ministry of Environment and Forests in 1985.

The constitutional provisions are backed by a number of laws—acts, rules, and notifications. The EPA (Environment Protection) Act, 1986 came into force soon after the Bhopal Gas Tragedy and is considered an umbrella legislation as it fills many gaps in the existing laws. Thereafter, a large number of laws came into existence as the problems began arising, for example, Handling and Management of Hazardous Waste Rules in 1989.

Legislation has passed the following environmental laws for prevention and control of environmental pollution such as **The Water (Prevention and Control of Pollution) Act, 1974** which provides for institutional structure for preventing and abating water pollution. It establishes standards for water quality and effluent. Polluting industries must seek permission to discharge waste into effluent bodies, **The Environment (Protection) Act, 1986** which authorizes the Central Government to protect and improve environmental quality, control and reduce pollution from all sources, and prohibit or restrict the setting and for operation of any industrial facility on environmental grounds, **The Indian Forest Act, 1927** and **Amendment, 1984,** was enacted to 'consolidate the law related to forest, the transit of forest produce, and the duty leviable on timber and other forest produce', **The Wildlife Protection Act, 1972** which provides for the protection of birds and animals and for all matters that are connected to it whether it be their habitat or the waterhole or the forests that sustain them, **The Forest (Conservation) Act, 1980** which provides for the protection of and the conservation of the forests, **The Air (Prevention and Control of Pollution) Act, 1981** which provides for the prevention, control and abatement of air pollution, to lay down the standards to maintain the quality of air etc.

There are also many laws relating to hazardous waste management such as Bio-Medical Wastes (Management and Handling) Rules, 1989 and these rules are applicable to all those persons who generate, collect, receive, store, transport, treat, dispose, or handle bio-medical wastes in any form, Hazardous Wastes

(Management and Handling) which contains contaminants and prevent accidents and limit their consequences on humans and the environment and provide persons working on the site with information, training and equipment necessary to ensure their safety.

Role of Judiciary in Environmental Preservation and Management

For the protection of environment from pollution, the Supreme Court of India and also various High Courts have taken innumerable measures in a series of their landmark judgements. Over the years, the apex court has been paying special attention for the protection of environment by giving effective directions to all the persons concerned with the matter by invoking in powers under the Article 32. An analysis of the various decisions of the Supreme Court reveals that the apex court exercised its writ jurisdiction when there was leakage of hazardous gases like chlorine from the Sri Ram Industries. In the past decade, the Supreme Court of India has shown respect and considerable concern to bring environmental awareness in the people and regarding this, there exist some leading cases (Anthony, 2005).

In the case of **M.C. Mehta** v. **Kamal Nath,*** Supreme Court declared that the public trust doctrine as developed by the ancient Roman Empire is a part of the law of the land and the natural resources which are meant for public use cannot be converted into private ownership. In the present case, large area on the bank of the river Beas which is the part of the protected forest has been given on a lease by the Himachal Pradesh Government purely for commercial purposes to the motels. River Beas which is a young and dynamic river and is considered to be ecologically fragile. This scenic and beautiful riverside should not have been permitted to be converted into private ownership and for chimerical gains on the notion being that the public has a right to expect certain lands and natural areas to retain the natural characteristics. Thus taking into consideration all these, The Supreme Court observed that "we have no hesitation in holding that the Himachal Pradesh Government committed patent breach of public trust by leasing the ecologically fragile land to the motel management". The court opined that "our legal system-based

* AIR, 1999 SC 2968.

on English common law—includes the public trust doctrine as part of its jurisprudence. The State is the trustee of all natural resources which are by nature meant for public use and enjoyment. Public at large is the beneficiary of the seashore, running waters, airs, forests and ecologically fragile lands. The State as a trustees under a legal duty to protect the natural resources. These resources meant for public use cannot be converted into private ownership" (www.answer.com, 2007).

Thus, holding the above view, the Supreme Court directed that "the doctrine of public trust as discussed by us in the judgment is a part of the law of the land" and cancelling the lease granted to the motel, set it aside and also asked the Government of Himachal Pradesh to take over the area and restore it to its original-natural conditions. Further the court asked the motel to pay compensation by way of cost for the environment and ecology of the area. The pollution caused by various constructions made by the motel in the riverbed and the banks of the River Beas has to be removed and reversed.*

Municipal Council, Ratlam v. Vedhichand

Brief Facts—In the present case the residents of a locality within limits of Ratlam Municipality tormented by Stench and public excretion by nearby slum dwellers moved the Magistrate under Section 133 of Cr.P.C. to require Municipality to do its duty towards the members of the public. The Magistrate gave directions to Municipality to draft a plan within six months for removing nuisance. In appeal, Sessions Court reversed the order. The High Court approved the order of Magistrate. In further appeal, the Supreme Court also affirmed the Magistrate's order.

Held—Where there existed a public nuisance in a locality due to open drains, heaps of dirt, pits and public excretion by humans for want of lavatories and consequential breeding of mosquitoes, the Court could require the Municipality under Section 133 of Cr.P.C. and in view of the Section 123 of the Municipalities Act to abate the nuisance by taking affirmative action on a time bound basis. When such order was given the Municipality could not take the plea that

* M.C. Mehta *vs.* Kamal Nath, (1977) 1 SCC 388 at p. 415.

notwithstanding the public nuisance, financial inability, validly exonerated it from statutory liability.

Code Operates Against Statutory Body—The Criminal Procedure Code operates against statutory bodies and other regardless of the cash in their coffers, even as human rights under Part III of the Constitution have to be respected by the State regardless of budgetary provision. Likewise, Section 123 of the Municipalities Act has no saving clause when the Municipal Council is penniless. Otherwise a profligate statutory body or pachydermic governmental agency may legally defy duties under the law by urging in self-defence a self-created bankruptcy or prevented expenditure budget. That cannot be.

Section 133 of Cr.P.C. is categoric and has a mandatory import. Although reads discretionary judicial discretion when facts for its exercise are presented. Therefore, when the Magistrate has, before him, information and evidence, which disclose the existence of a public nuisance and on the material placed, he considered that such unlawful obstruction or nuisance should be removed from any public place which may be lawfully used by the public, he shall act. Thus, this judicial power shall, passing through the procedural barrel, fire upon the obstruction or the nuisance, trigger by the jurisdictional facts. The Magistrate's responsibility under Section 133 of Cr.P.C. is to order removal of such nuisance within a time to fix in the order. This is a public duty implicit in the public power to exercise on behalf of the public and pursuant to a public proceeding. Failure to comply with the direction will be visited with the punishment contemplated By Section 188 I.P.C., therefore, the municipal and the other executive authority is bound by the order under Section 133 of Cr.P.C. and shall obey the direction, because disobedience, if it causes obstruction of allowance of injury to any person lawfully persuing there employment shall be punished with simple imprisonment of fine as prescribed in the section. The offence is aggravated if the disobedience tends to cause danger to human health or safety. The imperative of Section 133 of Cr.P.C. read with the punitive temple of Section 188 I.P.C., make the prohibitory act a mandatory duty.

Public Nuisance must be Checked—Public nuisance because of pollutants being discharged by big factories to the detriments of the poorer section, is the challenge to the 'social justice' component of the 'Rule of Law'. Likewise, the grievous failure of local authorities to provide basic amenity of the public conveniences drives the miserable slum-dwellers to ease in the States on the sly for a time and openly thereafter. In the following terms these slums and the survival thereon become a deeply entrenched part of the particular urban scenario, further magnifying the trauma of environmental pollution. A responsible Municipal Council constituted for the precise purpose of preserving public health and providing better finances cannot run away from its principal duties by pleading financial inability. Decency and dignity are non-negotiable facets of human rights; rather are a foremost obligation and charge on local self-governing bodies. Poor sanitation provision is another wastable instrument for creating Public Nuisance. Hence, providing drainage system though not pompous and attractive, but atleast in working condition and sufficient to meet the sanitation related needs of the people cannot be evaded if the municipality is to justify its efficiency.

Sachidanand Pandey *vs.* State of West Bengal

In this case the scope of judicial review in environmental cases was explained by the Apex Court.

Brief Facts—Where a group of citizens challenged the location of a hotel on the ground that the construction of hotel would interfere with the flight path of Migratory Birds. However, the construction of the Calcutta Taj Hotel was delayed for six months while administrative agencies and courts considered whether a proposed six-storey hotel would impede the flight pattern of migratory birds. The land on which the hotel was to be constructed formally belonged to the Alipore Zoological Garden and was put to important zoo related uses. The zoo directors had withdrawn their objections to the hotel after the Government promised them adjacent lands and relocation grants, the hoteliers agreed to reconstruct all displaced facilities on the adjacent land, at no expense to the zoo.

Scope of Judicial Review—The least that the court may do is to examine whether appropriate considerations are brought are borne in mind and irrelevancy excluded. In appropriate case the court may go further but how much further must depend on the circumstances of the case. The court may always give necessary directions. However the court will not attempt to nicely balance relevant considerations when the question involves nice balancing of relevant considerations, the court may feel justified in resigning itself to acceptance of a decision of the considered authorities.

The Supreme Court expressed the view that it is not merely another term planning case but a ringing statement of the court's duty to protect the environment. Here, the duly authorized construction of medium-sized five star hotel was alleged to interfere with the flight path of migratory birds and the court inquired extensively into this question. If the migratory birds are worthy of the court's attention of *"fortiori"*, the court's protection would extend to environmental issue of more direct positive and desirable impact on human beings.

Held—Whenever a problem of ecology is brought before the court, the court is bound to bear in mind the Article 48-A of the Constitution and Article 51-A (g)…. when the court is called upon to give effect to the directive principle and fundamental duty, the court is not to shrug its shoulders and say that priorities are a matter of policy and so it is a matter for policy making authority. The least that the court may do is to examine whether appropriate consideration are borne in mind and irrelevancies excluded. In appropriate cases, the court may go further, but how much further will depend upon circumstances of the case. The court may always give necessary direction. However, Court will not attempt nicely balance relevant consideration. When the question is evolved the nice balancing of relevant consideration, the court may feel justified in resigning itself to acceptance of the decisions of the concerned authority.

Development Allowed—Though the citizens of Indian cities are facing environmental urban problems, however, the hotel project was eventually permitted to proceed, not only because the hotelier had taken environmental values into account in fashioning the project but also because there obvious public benefit, viz.—increased revenues from tourism, and general upgrading and the beautification of the area.

Suggestions

Looking into the present perspective, there is a grave need to take up certain steps to retort this problem of environmental degradation and refabricise the measures which have been taken to overcome these problems. There is need to intensify cooperation between different levels of Government—local, regional and national and each should be made responsible to protect the environment. Cooperation between the public sector and the private sector must grow in designing solutions to environmental problems. "A thoughtful use and explantation of the available natural resource is rather a functional prerequisite. It is better to buy only pine, oak, ash or beech because they can be replaced. Non-renewable resources should rather be preserved and used in a sustainable limits only. Habitat destruction is one of the main reasons why many species face extinction. In many countries souvenirs made from rare wildlife are available. Hence, strict measures should be taken against buying shells, coral or things made from elephant ivory, rhino horn or cat skin etc. The destruction of wood land, pollution of rivers and ponds, the use of pesticides and herbicides have all contributed to the extinction of wildlife. Old and redundant clothes should be taken to the charity shops, so that some are sold on lower prices to the poor and others are returned to textile mills for recycling. When fossil fuels, i.e. oil, gas and coal are burnt to provide energy for lighting, cooking etc., besides releasing polluting gases, it consumes the already depleting natural resources. It becomes the duty of concientious citizens not to waste electricity and switch off lights when rooms are not in use. It is a well known fact that Methane is the most effective greenhouse gas, which makes the sites rancid." Hence, its release should be controlled. If we use chlorine-based bleach or detergents containing phosphates, we are contributing to water pollution, so try to buy 'environmentally-friendly' products. As we know that a lot of packaging, e.g. fast-food cartons, are polystyrene 'blown' with CFCs, an attempt should be made to avoid items packed with this polystyrene. Certain gases in the atmosphere, mainly carbon dioxide, methane and CFCs, act like the glass in a greenhouse allowing sunlight through to heat the Earth's surface but trapping some of the heat as it radiates back into space. Without this the Earth would be frozen and lifeless. However, owing to Man's activities, 'greenhouse gases' are building up in the atmosphere, causing a greater amount of heat to be reflected back to Earth. The result is an increase in average world temperatures and in the future this could lead to the melting of polar ice and flooding of cities world-wide besides more hurricanes etc. Moreover, on burning, coal, oil and gas releases a number of obnoxious gases such as oxides of nitrozen oxide, sulphur dioxide, carbon dioxide etc. Also car fumes produce carbon dioxide and nitrogen oxide. Though the use of the private owned cars in modern hectic era cannot be ignored, but as a matter of civil moral obligation, more and more use of public transport should be made. Ordinary people should be given training to recycle the available waste in households, which will help to control ecological degradation to a significant extent. We should stop this annihilation of our environment and give the coming generations a chance to live in clean environment which is free from pollution of all sorts. Though practising these rendered solutions is not so practically possible due to lack of individual environmental conscience in the citizens; still attempts can be made to mobilise them with the help of assistance of the Governmental as well as Non-Governmental bodies functioning for the cause of environmental protection, preservation and management.

REFERENCES

Anthony, M.J., *Landmark Judgements on Environment and Pollution;* Indian Social Institute; New Delhi; 2005; p. 5.

Jaiswal, P.S., *Environmental Law*: Pioneer Publications; Delhi; 2006; p. 79.

Padme, *Environmental Protection and the Law;* Gogia Law Agency; Hyderabad; 2006; p. 55.

Tiwari, A.K., *Environmental Laws in India;* Deep and Deep Publications Pvt. Ltd.; New Delhi; 2006; p. 24.

Tripathi, S.C., *Environmental Law;* Central Law Publications; Allahabad; 2005; p. 16.

Verma, S.B., *Environmental Law Pollution and Management;* Sital Offset Press; Jaipur; 2003; p. 34.

http/www.answers.com/Indian environmental law/25/11/2007.

15

Humanitarian Faces and Legal Hues of Aura

GIGIMON V.S.

Fortification of environment spills a bunch of problems for a nation. There are many reasons behind this rampant use of science and technology and it itself becomes one among the reasons for environmental degradation and threatening the existence of mankind. Justice V.R. Krishna Iyer has rightly said:

> *"Human survival is menaced by another equally homicidal missile euphemistically described as environmental pollution. If I may mint an odd expression, 'thanatalogy through technology' is the Frankenstein's monster that science and industry, by promising global progress, have created. If dehumanized industrialization, with all its profit-hungry vulgarity and its ecological insensitivity, invade Nature with enlightened resistance from society and persons or depletes all the resources of land, water and air, the crucification of humanity is a certainty and the resurrection of the race a lost possibility unless we begin the battle for human values against barbarity incorporated, right now. Today is right; Tomorrow may be too late... we have guided missiles and unguided men."*[1]

The terms like 'Environment', and 'Environmentalism', 'Environmentalists' have become common usage words for the developing nations. As the concern for the environmental protection is taking a central stage, discourse on environment and its related issue is a matter of concern for academic, research and governance.

The term environment has been derived from a French term[2], which means '*to surround*'. Thus etymologically implied meaning of environment is '*surrounding conditions, circumstances affecting people's life*'.[3] But the dictionary meaning of the term environment is '*the conditions, circumstances etc, affecting a person's life*'[4] or '*the natural world, within which people, animals, and plants live*'[5]. Thus any act that affects or pollutes the environment will affect the life of the person and those acts, which threaten the existence of human being, can be punishable as violative of Human Rights.

Many writers and jurists have attempted to give a clear definition to the term. Cadwell said that "it is a term that every one understands and no one is able to define".[6]

According to Gilpin, "from a scientific point of view, it is taken to mean everything that is physically external to the organism; organisms of course include human beings".[7]

According to Justice P.N. Bhagwati, environment refers to "the conditions within and around an organism, which affect the behaviour, growth and development, of life processes, directly or indirectly. It includes the conditions with which the organism interacts".[8]

The above stated definitions connote that there is every reason to believe that for the survival of living organisms, a balance has to be maintained between them. However, we are witnessing velocity and by moving with velocity, we are getting carried away with it. Thus, these comforts have already destroyed our fresh air, water and other ecosystems, which is essential for the future generations.

To overcome all these problems international treaties and conventions have laid down many norms. The first was laid down by the United Nations, in its International Conference[9] in the year 1972, in its agenda as "*to defend and improve the human environment for present and future generations has become an imperative goal for mankind*". It called upon all the Governments and people of the world to exert common efforts for the preservation and improvement of the Human Environment. This Conference at Stockholm became the turning point for the Environmental Jurisprudence. It laid down twenty-six principles, which are known as *Magna Carta* on Human Environment. The principles are categorized under the various heads:

(a) two proclamation rights[10];
(b) four concerning the conservation of resources[11];

(c) two on pollution[12];
(d) eight on developmental issues[13];
(e) nine on specific non-legal topics[14]; and
(f) one on State responsibility[15].

Another popular international effort for the cause of preservation of the environment is the Earth Summit.[16] The document produced at the Earth Summit has 40 chapters spread over 800 pages. In this Conference, 108 States participated. They together adopted three major agreements aimed at changing the traditional approach to development:

- Agenda 21—a comprehensive programme for global action in all areas of sustainable development;
- The Rio Declaration on Environment and Development—a series of principles defining the rights and responsibilities of States;
- The Statement of Forest Principles—a set of principles to underlie the sustainable management of forests worldwide.

Indian Parliament passed the 42nd Amendment to the Constitution in 1976[17] and incorporated specially two articles relating to protection and improvement of the environment. Thus, India became the first country in the world to leap forward for preventing this precious aura and have provisions in the Constitution. Despite the provisions in Constitution of India, the Parliament has enacted various laws governing the protection and preservation of the environment. For example, Water (Prevention and Control of Pollution) Act, 1974; Air (Prevention and Control of Pollution) Act, 1981; Environment (Protection) Act, 1986; Forest Conservation Act, 1980; National Highways Act, 1983; The Drug and Cosmetic Act, 1990; Motor Vehicle Act, 1988; Industries (Development and Regulation) Act, 1951; Wild Life Protection Act, 1972; etc. These are the specified legislations for the environmental protection. There are many other provisions made for the same in I.P.C.[18] and Cr.P.C.[19] for punishing the persons who can damage the environment, threatening the existence of mankind.

The Earth Summit succeeded in presenting new perspectives on economic progress. It was lauded as the beginning of a new era and its success would be measured by the implementation, locally, nationally and internationally, of its agreements. The Earth Summit proved colossus amongst all subsequent UN conferences, which have examined the relationship between human rights, population, social development, women and human settlements and the need for environmentally sustainable development. Thereafter, the World Conference on Human Rights, held in Vienna in 1993, emphasized the right of people to a healthy environment and the right to development, and controversial demands that had met with resistance from some Member States until the Earth Summit.

Then there was the General Assembly Resolution in 1972,[20] emphasizing the need for the active cooperation among the States in the field of human environment. Then, there was charter of Economic Rights and Duties of States, in 1974, which stressed on the protection, preservation and the enhancement of the environment for the present and the future generations. It basically encouraged and guided the States to undertake awareness programmes; rather inspired the Nations concerned not to undertake such economic activities which could further harm the environment of the area.

All these enactments contain provisions for the protection of environment by imposing it as a duty of every citizen as well as on the State, to protect, preserve, conserve and manage the environment. But these provisions are not all positive in nature and do not bring about 'Welfare Oriented Change Only'. It is not always true to say that, the laws pertaining to protection of environment violate human rights but there are provisions under certain laws which when exercised abridge the rights of persons in exercising their free will. For instance, in the Forest Conservation Act, 1980, in which the Central Government is the sole authority to give directions regarding the reserved forests and on such recommendations only the State Government can further make rules. By utilizing this power the State Government has made rules, which says that no person shall be allowed to enter into the forests or go out without the permission of the forest authorities. Similarly, the rules again say that any property lying in the reserved area belongs to the Government and no other use of forest land or any portion of it can be made for any non-forest purpose[21]. Here the rule is enacted to protect the forest, but the question arises about the rights of persons who were residing in these forests.

These tribal people were the protectors of these forests, the rights of the sons of the soil. By implementing these provisions the rights of these tribal people are infringed who were dependent on these forest for their livelihood.

On one pretext or the other, most of these environmental laws violate the rights of the people, the reason is that the Government is creating awareness of protecting the environment forgets about the human rights. These Acts hardly give any consideration to the living beings called human, the **human face** because their first concern was environment protection. For protecting and preserving the Human Rights, the courts have to interfere to provide appropriate remedy to the aggrieved, excavating their rights from Article 21[22].

The Supreme Court through a catena of decisions has opened many facets of Article 21 to protect the human rights with environment. In **Rural Litigation and Entitlement Kendra v. State of Uttar Pradesh**[23] the right to clean **environment as an integral** part of right to life.

Other conflicting areas where environmental law lacks human face are:

(a) Under the right to life drinking water is one of the fundamental requirements but simultaneously State is obliged to preserve the ground water or to restrict the use so that it can be prevented from being wasted.[24]

(b) State has an obligation to protect the forest but at the same time has to protect the tribal people whose livelihood is based on forest.

(c) State is bound to close down those industries, which are hazardous for environment, but at the same time has to keep in mind the livelihood of the people who work there.

The Supreme Court from time to time has been issuing guidelines for the protection of Human Rights and Environment through various case laws. The Supreme Court had taken into account the right to healthy environment along with the right to sustainable development and balanced them. This concept of right to healthy environment and sustainable development are the fundamental human rights implicit in the right to life, which has been constructed as such in many countries. The entire judicial construction by the Supreme Court and the High Courts also reveal the humanitarian approach to these environmental laws with the help of the public interests litigations. The Supreme Court of India was the first to develop the concept of right to healthy environment as a part of life under Article 21 of our Constitution.[25] This principle is now being adopted and followed by many countries.

But mere creation of the law for the stated purpose is not sufficient. It has to be seen that the conditions suitable for its execution are well equipped to digest the recently made implementation. For implementing these, many things have to be considered like the laws passed by the legislature and its institutionalization and internalization. The Supreme Court in **M.C. Mehta** vs. **Union of India**[26] accepted the Roman doctrine of Public Trust, which means that the State is the trusty of all the property which should be used for the public purposes and State should also restrict the private ownership in this field.

State being the guardian of individuals residing there has to protect and secure their rights. Moreover, with the emerging concept of welfare State, the State must reach at the minutest level. It should make laws, which do not impede the rights of persons, and if there is any such requirement for such law it should be implemented without doing much harm. Some accountability should also have to be included for the reason that if any wrong has been committed State should be made accountable for it.

For example, Narmada Valley Dam Project, a development project by the State for the development of Narmada Basin for irrigation, power generation and providing water for industries and domestic requirements for the State of Madhya Pradesh. But at the same time, the project involves devastation of human lives and biodiversity by inundating thousand acres of forests and agricultural land. The State wants these dams on the Narmada River in the name of national development. The people who suffer are the tribals who have been living therefrom time immemorial. Here what we want is the little humanitarian approach to these people. If they were to be relocated, will they survive there? will they be able to do the same work for the livelihood etc... These things have to be given consideration together with the development of the nation. We can not develop a nation by destroying many peoples' livelihood. Here in these circumstances, the accountability part is still not to be found. Who should be made accountable for these problems?

In such circumstances, public participation is another way to check the governmental action, which lacks human face. Equally, Environmental Impact Assessment plays a very pivotal role. It helps in quality decisions. They should also assess impacts and complete an environmental impact statement, which should be opened to public for their commentaries. Such kind of participation makes easier for the project proponent to take advantage of the information that citizens contribute concerning values, impacts, innovative solutions and alternatives.

Like-wise, for maintaining a balance between the Human Rights and the environmental protection, the draft recommendation of the report by Committee on the Environment Agricultural and Local and Regional Affairs[27] can be considered. The draft recommendation states that—

"Various *modus oprendies* suggested by Assembly to the member states [28]:

- Ensure appropriate protection of the life, physical integrity and health of persons by also taking account of the need for environment protection.
- Recognize a human right to healthy, viable and decent environment including the object obligation for States to protect the environment.
- Safeguard the individual rights set out in the Aarhus Conventions on Access to Information, Public Participation in Decision Making and Access to Justice in Environmental Matter.
- Harmonize their laws on environmental protection and safety.

Although both human rights protection and environmental protection are relatively well-developed areas of public policy, the relation between the two has not yet developed. As NGOs and policy makers have increasingly encountered situations, calls for the protection of environmental rights have deepened. Laws have their ostensible effect as despite of so many reported issues of human rights violation, no binding international agreement has had environmental rights and human rights as its primary spotlight. In addition, the issue continues to suffer from inattentiveness due to the fact that it fails to fit neatly within the agenda of either the human rights movement or the environmental movement.

There is a prominent saying that "prevent and prepare is always better than repent and repair". This saying should be well justified. Environmental right, as a part of human rights law, now stands substantiated through judicial precedents and other sources of international law as "Right to Life' is an integral part of Human Rights Law. Even after having all these recommendations, precedents and legislations, still people fight for their basic human rights. Now its high time to find a permanent solution for this fault. The questions that haunt our mind is where we failed, where the law proved ineffective and what remedies can we make out for solving these problems for a human faces in environmental protection laws.

NOTES

1. Justice V.R. Krishna Iyer, *Environmnental Pollution and Legislative Solutions,* 1 (1984).
2. French term '*environmer*' and Latin term '*in-viron*'.
3. P.H. Collins, *Dictionary of Ecology and the Environment,* 62 (1990).
4. Oxford Advanced Learner's Dictionary, p. 387, (5th Edn. Oxford, 1996).
5. Microsoft Encarta Dictionary, 2004.
6. Cadwell, *International Environmental Policy and Law,* (Derham 1980).
7. A. Gilpin, *Dictionary of Environmental Terms,* 51 (Queens Land University Press).
8. Justice P. N. Bhagwati, The Crucial Conditions, in Survey of the Environment, 1991, p. 165 (The Hindu).
9. UN International Conference on Human Environment at Stockholm, from 5th to 16th June, 1972.
10. Principles 1 and 21of *ibid.*
11. Principles 2 to 5 of *ibid.*
12. Principles 6 and 7 of *ibid.*
13. Principles 8 to 15 of *ibid.*
14. Principles 16, 20, 23 and 26 of *ibid.*
15. Principle 22 of *ibid.*
16. United Nations Conference on Environment and Sustainable Development 1992 at Rio-de-Jenerio.

17. The Constitution (42nd Amendment) Act, 1976 incorporated Articles 48-A and 51-A to the Constitution of India.
18. Chapter XIV Sections 268 to 294A providing penal provisions to punish the person who pollutes the environment.
19. Chapter X Sections 133 to 146 and Chapter XI Section 144.
20. General Assembly's Resolution of 5th December 1972, designating June 5th as World Environmental Day.
21. Section 2 of the Forest (Conservation) Act, 1980.
22. In the year 1984, the Supreme Court gave the judgment that Right to Life also includes right to pollution free environment in the case *Bhandua Mukti Morcha* v. *Union of India,* 1984 (3) SCC 161.
23. AIR 1985 SC 652; See also, *Subash Kumar* v. *State of Bihar,* AIR 1991 SC 420.
24. The need for management of groundwater resources was recognized by the High Court of Kerala in a public interest litigation filed by the local islanders. *Attakoya Thangal* v. *Union of India,* 1990 (1) KLT 580.
25. *Bhandua Mukti Morcha* v. *Union of India,* 1984 (3) SCC 161.
26. 1997 (1) SCC 388.
27. Environment Agricultural and Local and Regional Affairs held at New Delhi on 16th April 2003.
28. Point 9 of *ibid.*

Index